CAMBRIDGE CHECKPOINT

ASPIRE
SUCCEED
PROGRESS

Complete Biology for Cambridge Secondary 1

Pam Large

WORKBOOK

Oxford excellence for Cambridge Secondary 1

OXFORD

OXFORD
UNIVERSITY PRESS

Great Clarendon Street, Oxford OX2 6DP

Oxford University Press is a department of the University of Oxford.
It furthers the University's objective of excellence in research, scholarship,
and education by publishing worldwide in

Oxford New York

Auckland Cape Town Dar es Salaam Hong Kong Karachi
Kuala Lumpur Madrid Melbourne Mexico City Nairobi
New Delhi Shanghai Taipei Toronto

With offices in

Argentina Austria Brazil Chile Czech Republic France Greece
Guatemala Hungary Italy Japan Poland Portugal Singapore
South Korea Switzerland Thailand Turkey Ukraine Vietnam

First published in 2013

British Library Cataloguing in Publication Data

Data available

ISBN 978-0-19-839022-0

20 19 18 17 16 15 14

Printed in China by LEO

Acknowledgments

®IGCSE is the registered trademark of Cambridge International Examinations.

The publisher would like to thank Cambridge International Examinations for their kind permission to reproduce past paper questions.

Cambridge International Examinations bears no responsibility for the example answers to questions taken from its past question papers which are contained in this publication.

Cover photo: Eduardo Rivero / Shutterstock

Artwork by: Q2A Media and Erwin Haya

Welcome to your **Complete Biology for Cambridge Secondary 1** workbook.

This workbook accompanies the Student Book and includes one page of questions for every two pages of the Student Book. Each question page includes several types of question.

- Some questions ask you to choose words to complete sentences. These questions will help you to learn and remember key facts about the topic.
- Other questions ask you to identify statements as true or false, or put statements in the correct order. Some of these questions are testing your knowledge, others are asking you to apply what you know to a new situation.
- There are many questions that ask you to interpret data from investigations, or information from other sources. When you answer these questions, you will be practising important science skills, as well as preparing for the Cambridge Checkpoint test.
- Some pages include comprehension questions. They ask you to read some information, and then answer questions about it. Many of these questions will help you develop skills of evaluation.
- Most pages have an extension box. Some of these questions will help you to extend and develop your science skills. Many others go beyond Cambridge Secondary 1 Science. They include content equivalent to Cambridge IGCSE® level. All the extension questions are designed to challenge you, and make you think hard. There aren't any spaces for your answers to these extension questions, so you'll need to work on a separate sheet of paper.

This workbook has other features to help you succeed in Cambridge Checkpoint and eventually Cambridge IGCSE:

- The glossary explains the meanings of important science words. It includes all the **bold** words in the student book, and others.
- The Cambridge Checkpoint-style questions near the back of the book are excellent practice for the Cambridge Checkpoint test. These have been especially written by the author to provide lots of practice before your exam.
- The Cambridge IGCSE questions show you what you are aiming for. These have been taken from real past papers. Give them a try!

I wish you every success in science, and hope you enjoy the workbook.

Table of contents

Stage 7

Stage 8

1.1 Leaves, stems, and roots

1 The diagram shows a simple flowering plant. Label its four main parts.

2 Read the following paragraph and fill in the gaps with words from the box below. Use each word once, more than once, or not at all.

Flowering plants have three organs all year round:

roots, ... and .... They also develop .... when they are ready to .... Each plant .... has a different function, but they work together to keep the plant alive. Roots take up water and .... from the soil. Leaves absorb energy from sunlight and make .... A stem holds up the .... and transports .... and .... up from the roots.

flowers	food	leaves	organ	water	stems	minerals	reproduce

3 Write each of the jobs below in the correct column of the table.

allows reproduction **absorbs light** **takes in minerals** **makes food**
holds plant in place **provides support** **produces seeds** **takes in water**

Leaf	Stem	Root	Flower

4 The structure of each organ suits its **function**. Name the plant organ that is:

a Wide and thin to absorb a lot of light. ....

b Tall and strong to provide a lot of support. ....

c Highly branched to spread through a large volume of soil. ....

5 Write ***T*** next to the statements that are true. Write ***F*** next to the statements that are false. Then write corrected versions of the statements that are false.

a Stems carry food from the roots to the leaves.

b The parts of a plant that make food are usually green.

c Roots are the only organs in a plant that need water.

....

....

E

The main function of a stem is to support leaves and flowers. The swollen, green stems of cacti have extra functions. Describe one of these extra functions and explain how it helps cacti survive in the desert.

1.2 Questions, evidence and explanations

1 Read the following paragraph and fill in the gaps with words from the box below. Use each word once, more than once, or not at all.

Scientists ask .... to find out about the world around them. They try to .... these questions by suggesting possible .... Scientists also design .... to collect evidence. If enough .... supports an explanation, it is usually .... by other scientists.

evidence answer explanations questions investigations accepted

2 Draw a line to link each piece of evidence to the explanation it supports.

Evidence
A growing plant takes in a lot of water.
A growing plant gains more mass than soil loses.
Seedlings stop growing if their leaves are cut off.
Growing plants take a gas out of the air.
Plants do not grow well in pure rainwater.
Leaves, stems and roots have tubes running through them.

Scientist's explanation
A plant's leaves make the food it needs.
Plants need small amounts of minerals to grow, and they get them from soil.
Plants do not get their food from the soil they grow in.
Plants are made from water.
Tubes carry water from roots to leaves and food from leaves to roots.
Leaves use a gas from the air to make food.

3 Look at the evidence in the table.

Number of leaves on plant	Extra growth after 2 weeks (cm)
0	0
2	3.0
8	12.5

a Suggest what question the scientist was trying to answer.

....

....

b What does the evidence show?

....

....

c Suggest an explanation for the results.

....

....

....

E

A scientist wants to test the explanation that plants use water to make their own food. If the explanation is correct, plants should grow faster if they are given more water.

a Describe how she could collect evidence to test this idea.

b Predict what results she would obtain and give reasons for your answer.

2.1 The human skeleton

1 The diagram shows a human leg bone and the joints at each end of it (A and B). Match each of the facts below to the correct joint (write them in the correct column).

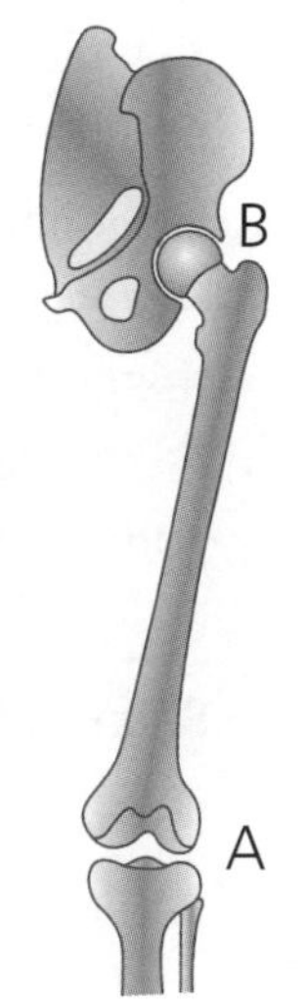

a hinge joint   **a ball and socket joint**   **connects leg to hip**
lets leg swing freely   **lets leg bend and straighten**   **found in the knee**

Joint A	Joint B

2 Each joint contains four different tissues. Link each tissue to the correct role with a line.

Tissue
Ligament
Cartilage
Bone
Synovial fluid

Role in the joint
Prevents the end of bones from banging together in a joint.
Provides support and can be pulled around by muscles to move your arms and legs.
Holds bones together but lets them swing freely.
Lubricates joints so bones can slide over each other smoothly.

3 Read statements a–e. Write ***T*** next to the true statements and ***F*** next to the false ones. Then write corrected versions of the statements that are false.

a Your skeleton protects delicate organs.

b Your backbone prevents damage to your heart and lungs.

c Your backbone is one long bone that runs down your back.

d All your bones are joined together by hinge joints.

e Joints allow your arms and legs to swivel around or bend.

Corrected versions of false statements:

....

....

....

E

The diagram below compares a normal joint with an arthritic joint.

a Name tissue A and fluid B.

b Describe how the arthritic joint differs from a normal joint.

c Explain how these differences affect the joint.

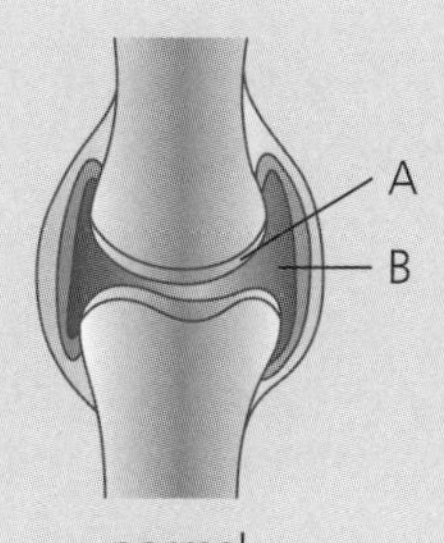

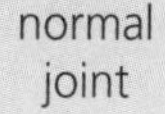
normal joint

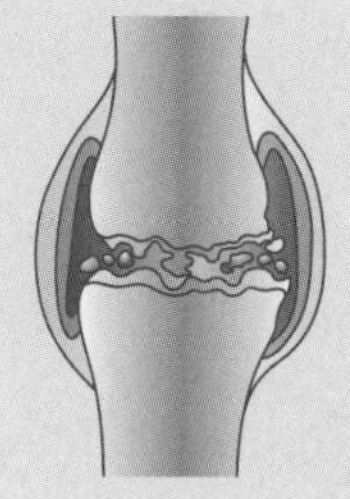
arthritic joint

1 This model explains how arm muscles work. Highlight the correct word in each **bold** pair.

In the model arm, the rulers represent **bones / muscles** and the strings represent antagonistic **ligaments / muscles**. The bent paperclips between them represent **tendons / joints**. When the string attached to A is pulled the model arm **bends / straightens**. When the string attached to B is pulled the arm **bends / straightens**.

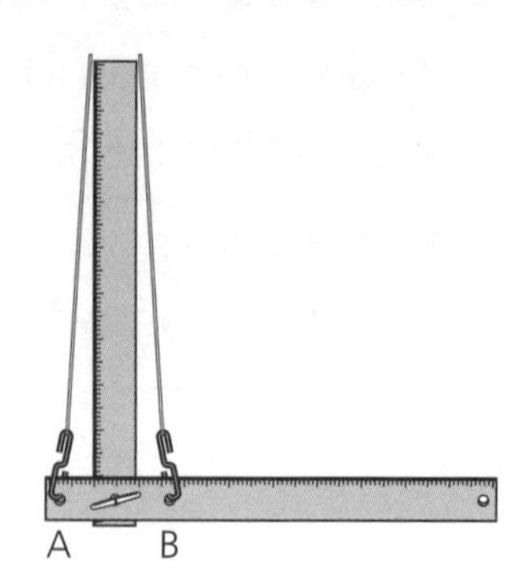

2 Read statements a-d. Write ***T*** next to the true statements and ***F*** next to the false ones. Then write corrected versions of the statements that are false.

a Tendons hold bones together.

b Muscles get shorter when they contract.

c Muscles push on bones to make you move.

d Most muscles are arranged in antagonistic pairs.

....

3 Four students write about antagonistic muscles.

Leah Most movements are caused by antagonistic muscles.

Acho There are pairs of antagonistic muscles in your arms and legs.

Karis When one muscle contracts to pull a bone, the antagonistic muscle relaxes.

Mikayla Muscles work in pairs because they can only pull bones in one direction.

a Give the name of the student who explained ***how*** they work. ....

b Name the student who explained ***why*** they work like this. ....

4 Diagram A shows the muscles that bend your arm.
Add the muscles that straighten your arm to diagram B.

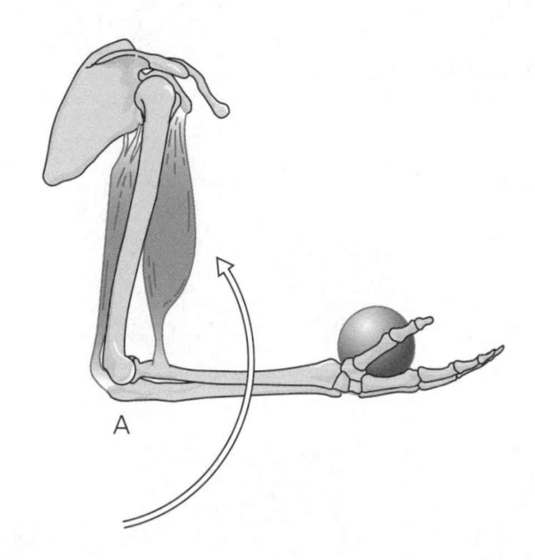

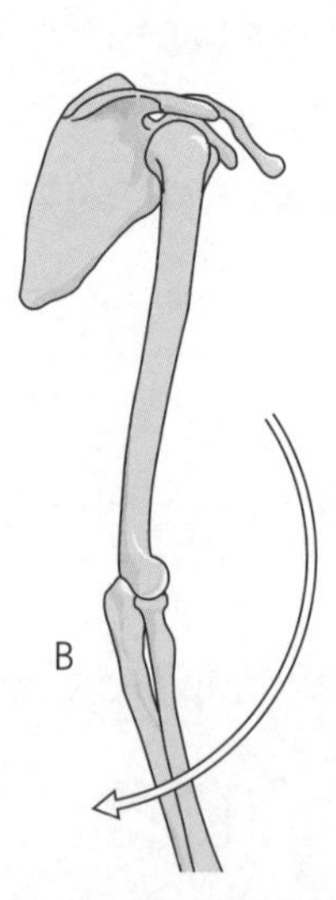

E

Carla sometimes has fits. During a fit, her arms and legs move rapidly in random directions. The fits are caused by uncontrolled activity in her brain. Explain how this brain activity could affect the muscles in her arms and legs.

2.3 Organ systems

1 The diagrams show four important organs.

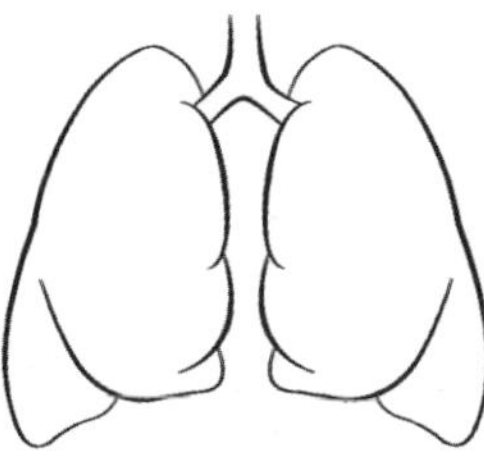

lungs

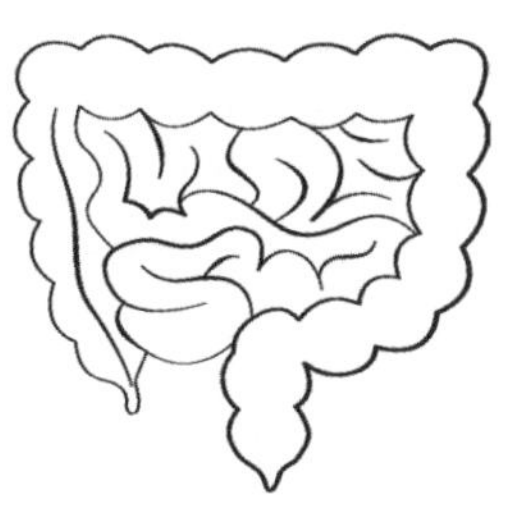
intestines

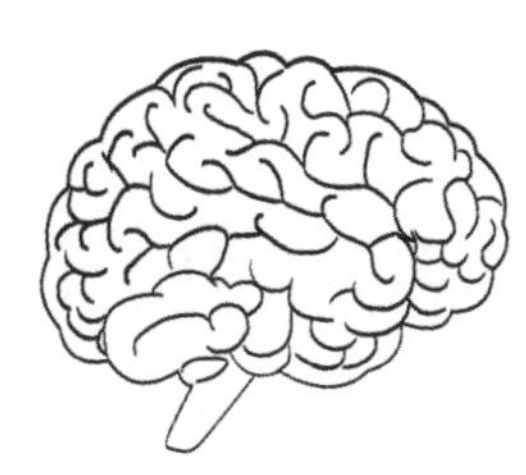
brain

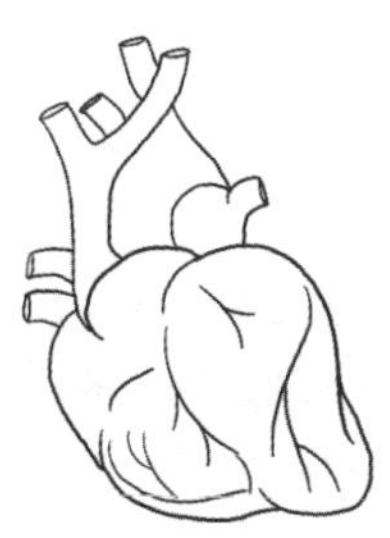
heart

Name the organ system each of these organs belongs to.

a Lungs ....

b Intestines ....

c Brain ....

d Heart ....

2 Draw lines to link each organ system to the correct role.

Organ system
Skeletal system
Muscular system
Nervous system
Respiratory system
Digestive system

Role in the body
Senses your surroundings and controls your actions.
Provides support and protection. Allows movement.
Breaks down large particles in food so they can get into your blood.
Contracts and pulls on bones to cause movement.
Carries out gas exchange. Adds oxygen to your blood and removes carbon dioxide.

3 Name the main organ system responsible for actions a to e.

a You spot a chocolate bar in the cupboard. ....

b Glucose particles from food enter your blood. ....

c Oxygen enters your blood and carbon dioxide leaves it. ....

d You bang your head but your brain isn't damaged. ....

e You move your arm to pick up your pen. ....

Put these statements into the correct order to describe how the nervous system works.

a Electrical messages travel along nerves to your brain.

b You turn your head towards the loud sound.

c Electrical messages travel along nerves to your muscles.

d Your ears detect the sound.

e There is a large bang behind you.

f Your brain interprets these signals as a loud sound.

Emran is running. Write a paragraph to explain how his nervous, skeletal and muscular systems are working together to make him run.

2.4 The circulatory system

1 The diagram shows how the heart pumps blood around the circulatory system.

a Label the heart and lungs.

b Label the artery that carries blood to capillaries all over the body.

c Label the vein that carries blood from the lungs to the heart.

d Colour the blood moving away from the lungs red.

e Colour the blood moving towards the lungs blue.

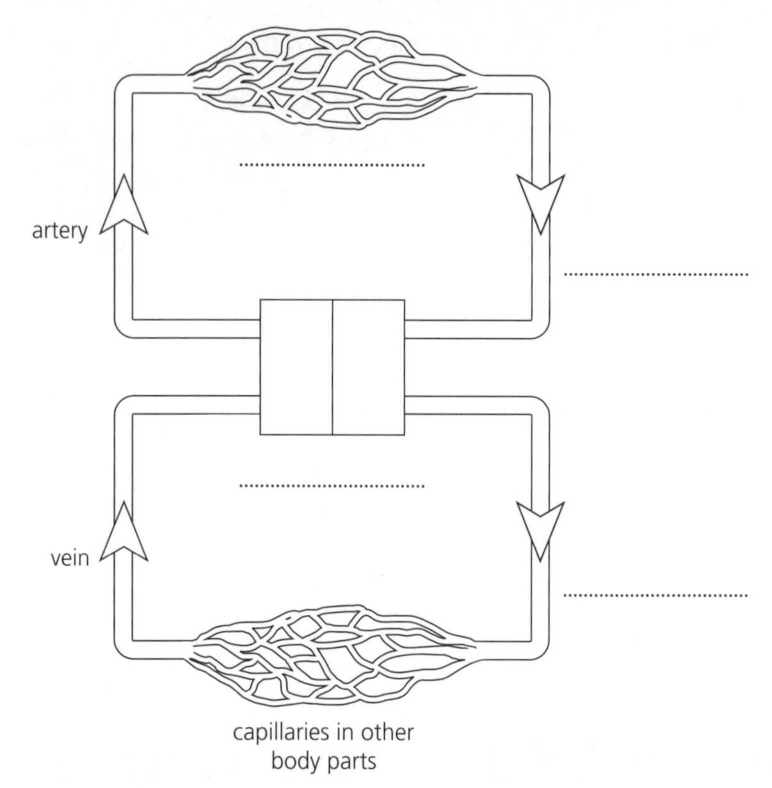

2 Complete these sentences about respiration using the words from the box below.

Respiration is a chemical ……………………. . It releases energy by reacting glucose with ……………………. . Blood collects glucose from your …………………... system and oxygen from your …………………... system. It carries these substances to every part of your body. You die if respiration stops because you run out of …………………… .

respiratory	digestive	oxygen	reaction	energy

3 Draw lines to match each of the roles below to the correct type of blood vessel.

Blood vessel
Artery
Vein
Capillary

Role
Thin-walled to let gases in and out.
Carries blood away from the heart.
Returns blood to the heart.

E

Heart muscles are too thick to get glucose and oxygen from the blood inside the heart. They take them from capillaries that run through the heart muscle. If the blood supply to these capillaries is cut off, the heart muscles stop working and cause a heart attack.

Explain why heart muscle stops working when its blood supply is cut off.

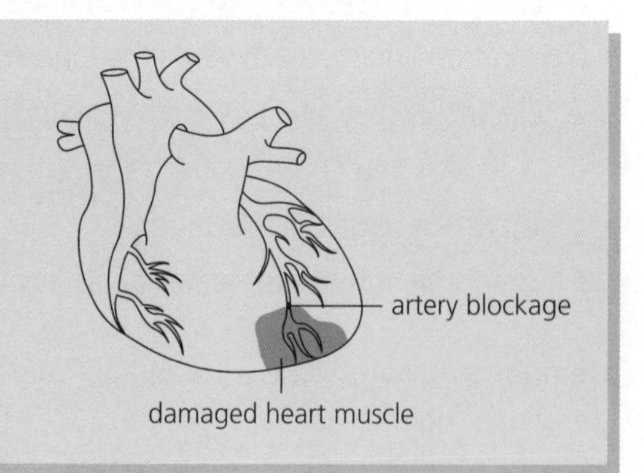

2.5 Studying the human body

1 Complete these sentences using the words in the box below.

Many different scientists study the human .... Their work helps us to stay .... , recover from illness or .... , or improve our .... Each type of scientist has specialist .... and skills.

lives	healthy	knowledge	body	disease

2 Use lines to match each scientist to the job that they do.

Scientist
Haematologist
Optometrist
Neuroscientist
Dietician
Prosthetic limb developer

Job
Examines your eyes to check for signs of disease and poor vision.
Examines blood to help doctors diagnose illnesses.
Recommends how we could improve our health by changing what we eat.
Makes replacements for missing hands, arms and legs.
Studies how the nervous system controls our bodies.

3 Choose the best type of scientist to help each of these students.

a Rina has been ill. A good diet could restore her health. ....

b Shanti needs to find out whether she needs glasses. ....

c Felicitas lost her foot in an accident but wants to walk again. ....

d Tarik feels tired all the time and his doctor does not know why. ....

e Kamran is depressed. He feels sad for no reason. ....

E

Neuroscientists want to find out whether depression makes patients more likely to:

a eat less nutritious food

b become less physically fit

c suffer from more illnesses

Choose three types of scientists who could help with their research and give reasons for your choices.

2.6 Extending lives

1 Use the words below to complete the following paragraph.

If you receive an organ .... , you need to take special drugs to avoid .... The organ transplanted most often is the .... We have .... kidneys. They remove .... from the blood and make .... We only need one kidney to stay healthy, so a .... donor can give one away.

rejection	live	waste	transplant	urine	kidney	two

2 Use lines to match each word to the correct definition.

Word
Transplant
Kidney
Urine
Scaffold
Reject

Definition
The organ that cleans your blood and makes urine.
A waste product made by your kidneys and stored in your bladder.
Move an organ from one person to another.
Fail to accept a transplanted organ.
What tissues are grown on to build body parts like ears.

3 Write ***T*** after the true statements and ***F*** after the false ones.

a An organ can be kept alive for a short time outside the body.

b Organs can only be transplanted from close relatives.

c After a transplant, special medicines need to be taken for a few weeks to prevent rejection.

d Organs grown from a patient's own tissues are never rejected.

4 Read the information in the box. Then answer the questions below.

> Lives can be extended by replacing faulty organs. Waiting lists for organ transplants are growing longer. Scientists want to solve this problem. They are trying to grow new organs from people's own tissues. They can already make simple body parts like windpipes and bladders. Organs like hearts and lungs will be more difficult to make.

a Suggest why more organ transplants will be needed in the future.

....

....

....

b Explain why a heart will be harder to grow than a bladder.

....

....

....

3.1 The characteristics of living things

1 Draw lines to match each word to the correct definition.

Word
Respiration
Sensitivity
Excretion

Definition
The removal of waste products from a living thing.
A chemical reaction that releases energy inside living things.
The ability to detect chemicals, light, heat, pressure or sound.

2 Tick the statements that apply to **all** living things.

- **a** They sense their surroundings.
- **b** They can only store waste products for a short time.
- **c** They are able to make their own food.
- **d** They release energy using chemical reactions.
- **e** They are able to produce offspring.
- **f** They are capable of movement.
- **g** They increase in size during their lifetime.
- **h** They must feed their offspring.

3 Living things have three ways of getting nutrients:

- **i** They make their own nutrients using a gas from the air and water.
- **ii** They obtain their nutrients by eating plants and other animals.
- **iii** They absorb nutrients from their surroundings.

Decide whether the organisms below use method i, ii or iii.

tree

fungus

snake

krill

house plant

a ……… **b** ……… **c** ……… **d** ……… **e** ………

4 Write three pieces of evidence that support the idea that trees are living things.

…………………………………………………………………………………………

…………………………………………………………………………………………

…………………………………………………………………………………………

E

Life on other planets may be too small to see. There may be tiny living things hidden in the soil or rocks. To detect life, scientists look for evidence of respiration.

Suggest two pieces of evidence that could show that respiration was occurring.

1 There are four types of micro-organism: fungi, protozoa, algae and bacteria. Identify the type in each of the following images. Each type may appear more than once or not at all.

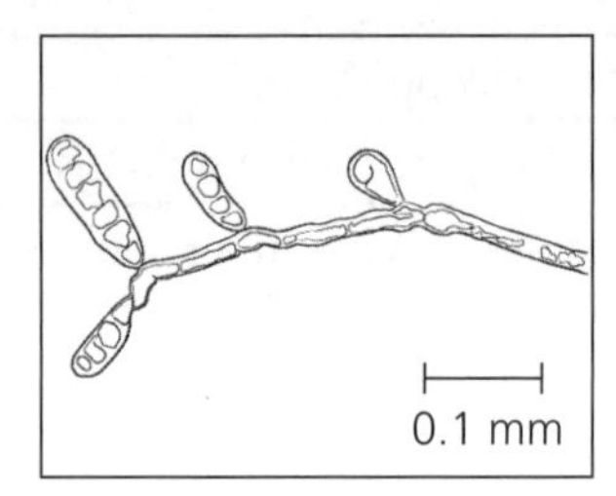

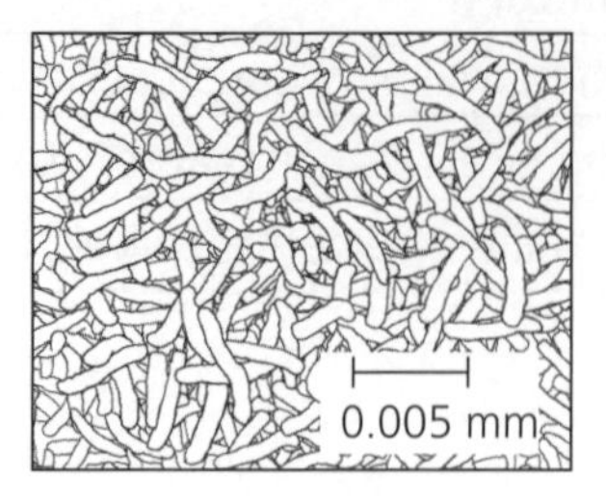

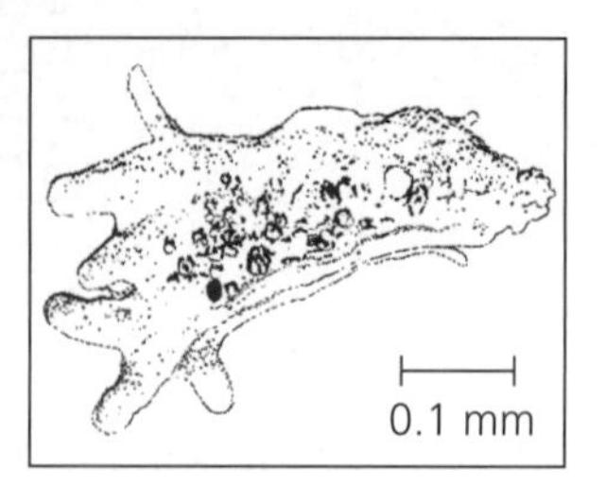

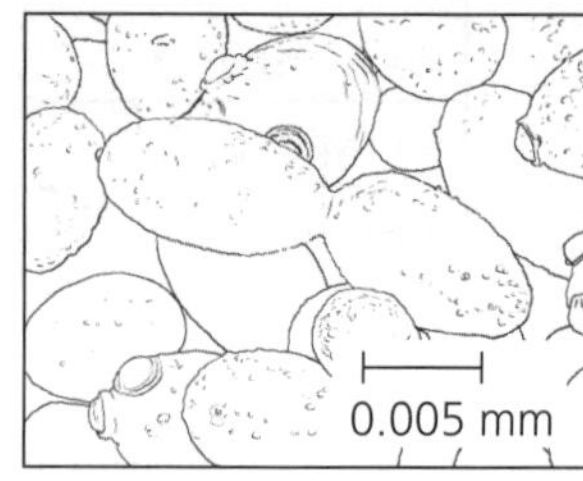

a .... b .... c .... d ....

2 Read the statements below and identify each type of micro-organism.

a It is made of long, thin threads and takes nutrients from its surroundings. It is a: ....

b They are very small and reproduce rapidly by splitting in two. They are: ....

c They are larger than bacteria and reproduce by budding. They are: ....

d It moves around in water and cannot make its own food. It is a: ....

e They can be small and round, or form long green strands. They are: ....

3 Write the letter of each statement in the correct part of the Venn diagram. For example: statement **a** is true for light microscopes and TEM microscopes so we write **a** where these circles overlap.

a The specimen slice must be very thin.

b It magnifies the object up to 1000 times.

c It can magnify the object more than 1000 times.

d The images can be coloured artificially.

e It shows the surface of specimens.

f Light passes through the specimen to make a magnified image.

g Used to look at organisms that are too small to see.

h Uses electrons to produce magnified images.

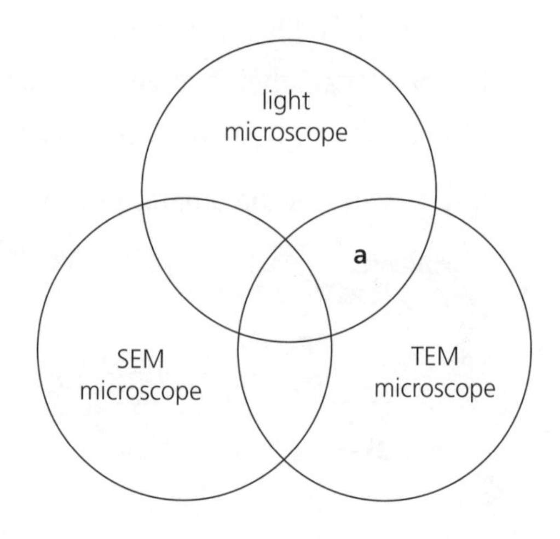

E

These are virus particles.

The particles cannot move but they can be carried around in the air. They can enter your lungs when you breathe. The virus makes you use nutrients and energy to produce more virus particles. The process damages your tissues and makes you feel ill. The new particles escape when you sneeze and infect other people.

a Write one fact that *suggests* viruses are living things.

b Explain how you can tell they are *not* living things.

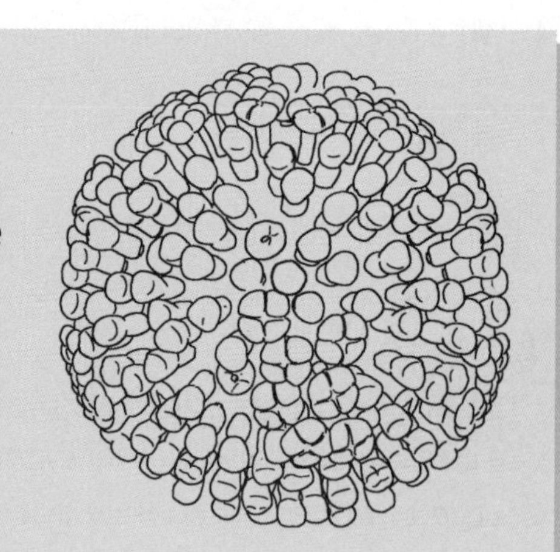

3.3 Louis Pasteur

1 Draw lines to match each word to the correct definition.

Word
Ferment
Alcohol
Pasteurisation
Organic matter
Lactic acid

Definition
A sour chemical some bacteria produce during respiration.
Materials made by living things.
Convert sugar into alcohol and carbon dioxide.
Heating food or drink for long enough to destroy most of the micro-organisms it contains.
A chemical produced by yeast during respiration.

2 Decide whether the following statements are about yeast, bacteria, or both.

a Used in wine and beer production ....

b Use sugar for respiration ....

c Used to produce alcohol ....

d Can produce lactic acid as a waste product ....

e Can make milk turn sour ....

f Can break down organic matter ....

3 The sentences below describe how milk turns sour, but they are in the wrong order. Write the steps in the process in the correct order.

a Lactic acid is produced.

b Milk is left in a warm place.

c Bacteria reproduce rapidly.

d The milk turns sour.

e Bacteria use sugar in the milk for respiration.

The correct order is ....

E

'Bioethanol' is alcohol produced as a fuel for cars. It can be made from most plant materials. There are three main steps in the production process: the plant material is broken down and mixed with water; yeasts are added to carry out fermentation; and the mixture is heated to separate the alcohol from the water.

a Explain what happens during the fermentation stage.

b Suggest why bacteria need to be kept out of the fermenting mixture.

3.4 Testing predictions

1 Pasteur used flasks of nutrients to investigate fermentation.

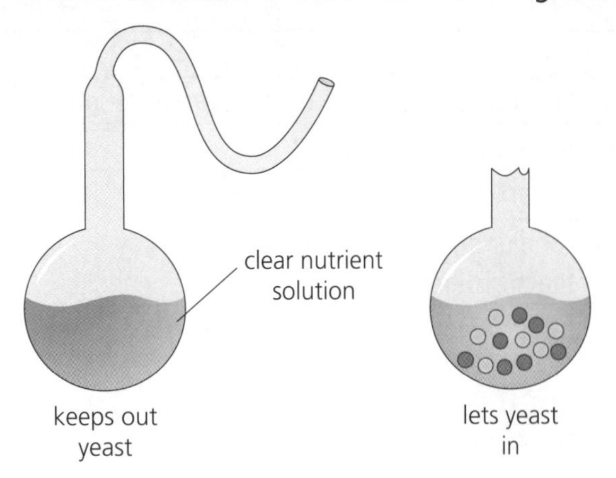

Draw lines to match what Pasteur does with a stage of developing an explanation.

What Pasteur does
Says: When yeast land in liquids full of nutrients they grow and reproduce, and make them ferment.
Thinks: If I keep yeast out of a nutrient solution it will not go cloudy because it will not ferment.
Places nutrient solutions in flasks with S-shaped necks, and boils them to destroy any micro-organisms present.
Notices that nutrient solutions do not go cloudy in flasks with S-shaped necks.
Breaks the neck of one of the flasks and observes that the nutrient solution begins to ferment and turn cloudy.

Stage of developing an explanation
Make a prediction.
Suggest an explanation.
Review the evidence.
Collect extra evidence.
Test the explanation.

2 In the table below, tick one box next to each statement to show whether the statement describes evidence, or whether the statement is an explanation.

Statement	Evidence	Explanation
When a nutrient solution ferments, bubbles appear.		
Yeast use nutrients for respiration and make carbon dioxide.		
Nutrient solutions turn cloudy when they ferment.		
If yeast fall into nutrient solutions, they grow and reproduce.		
Boiled solutions stay clear in flasks with S-shaped necks.		
Boiling destroys micro-organisms.		
S-shaped necks stop micro-organisms falling into flasks.		
Boiled solutions go cloudy in flasks with broken necks.		
Yeast from the air can fall into flasks with broken necks.		

E

Suggest how these predictions could be tested.

a Bubbles of carbon dioxide will appear faster if more yeast is added to fruit juice.

b Bacteria will produce lactic acid more quickly at higher temperatures.

1 Read the following paragraph and fill in the gaps with words from the box below. Each word may be used once, more than once, or not at all.

Bread is made by baking ………………., which is a mixture of ………………. and water. To make bread rise, ………………. can be added to the mixture. It uses nutrients from the flour for ………………. and releases bubbles of ……………… ………………. gas. The gas bubbles make the bread ………………. and spongy.

respiration	yeast	dough	soft	carbon dioxide	flour

2 Two bottles of milk were kept in a warm place. One had extra bacteria added to it. The pH of each bottle of milk was measured at the start and after 4 hours.

Look at the evidence in the bar chart.

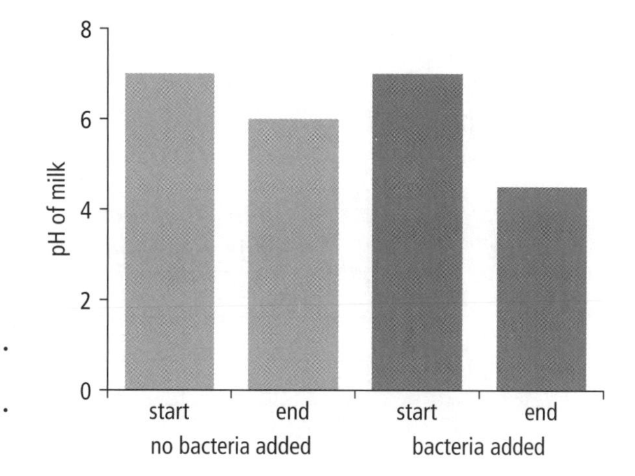

a How was the milk affected when extra bacteria were added?

…………………………………………

…………………………………………

b Explain how bacteria change the sugars in milk.

…………………………………………………………………………………

…………………………………………………………………………………

…………………………………………………………………………………

…………………………………………………………………………………

E

When yeast is added to sugar solutions they produce carbon dioxide. Nadia counted how many bubbles of gas were produced per minute at different temperatures.

Temperature (°C)	Bubbles per minute
30	60
35	78
40	100
45	52
50	5

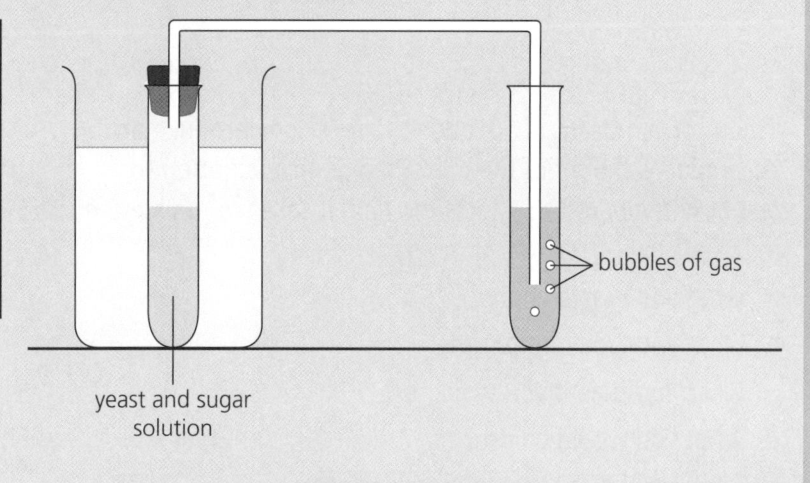

a Display these results on a graph and draw a curve through the points.

b Explain the shape of the graph.

3.6 Planning investigations

1 Read the following paragraph and fill the gaps with words from the box below. Each word may be used once, more than once, or not at all.

To plan an investigation we need to decide what to .... to answer a

........................, and what to .... to show the .... of the change.

Other variables that could affect the results need to be ...., which means

kept the ....

controlled	same	measure	question	effect	change

2 A student placed yeast in four flasks of warm water. She added different nutrients to each flask. Then she placed a balloon over the neck of each flask. After 30 minutes she measured the diameter of each balloon.

Nutrient added	Diameter of balloon (cm)
glucose	9.5
lactose	2
sucrose	5
fructose	4

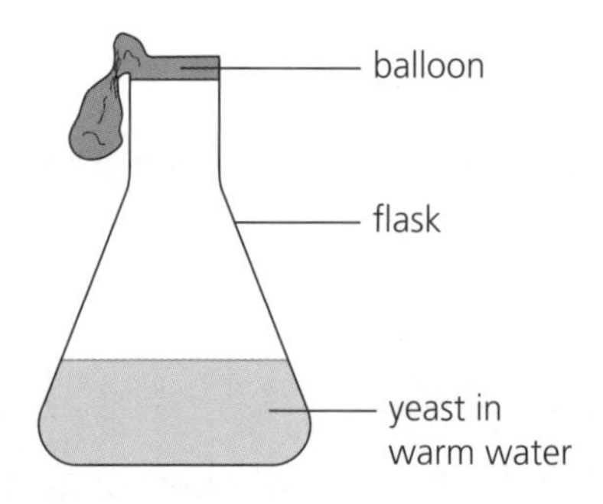

a Which variable did the student change? ....

b Which variable did she measure? ....

c List three variables that needed to be kept the same ....

....

....

d Name the gas that fills the balloons. ....

e Name the reaction that produces the gas. ....

f Name the nutrient that lets yeast respire fastest ....

E

When certain bacteria are added to milk they produce acid. This reduces the pH of the milk. Plan an investigation to test how temperature affects the time it takes to reduce the milk's pH to 5. Describe:

a what you would change

b what you would measure

c what equipment you would use

d what you would control

e what results you would expect to obtain.

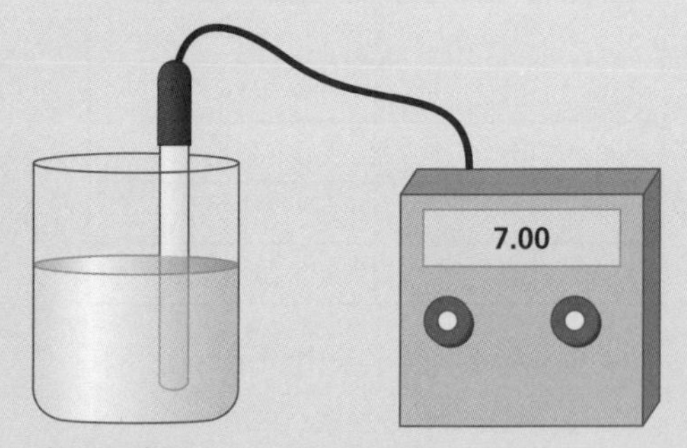

The reading on a pH meter drops as the amount of acid increases.

3.7 Harmful micro-organisms

1 Draw lines to match each infectious disease to the correct cause.

Infectious disease
Athlete's foot
Typhoid
Malaria
Flu
Hepatitis

Cause
Bacteria that infect the digestive system
Protozoa injected into the bloodstream by mosquitoes
A fungus that grows on skin
A virus that infects the liver
A virus that infects your lungs

2 Use their symptoms to decide which diseases these patients have.

a I have a headache and a fever. I have had diarrhoea for the past four days. ....

b I have a fever and bad stomach pains. I feel tired all the time. ....

c The skin between my big toe and second toe is very itchy. ....

d I have a fever and my muscles hurt. I get sudden chills. ....

e My head is very itchy and chunks of my hair have fallen out. ....

3 Write ***T*** next to the statements that are true. Write ***F*** next to the statements that are false.

a Hepatitis C can spread from person to person in infected water.

b Mosquitoes spread the bacteria that cause malaria.

c Some viruses can be spread by sharing contaminated needles.

d Typhoid is spread by a vector.

e Colds and flu are spread by virus particles in the air.

f Frequent hand-washing can reduce the spread of infectious diseases.

4 Malaria is common in many parts of Africa, Asia and South America.
Suggest two reasons why the disease has been so hard to eradicate.

....

....

....

....

E

The pathogen that causes typhoid is present in each victim's faeces.
Suggest three things a government could do to reduce the spread of this disease.

People infected with hepatitis C only develop symptoms after many years.
Suggest why this makes it harder to control the spread of the disease.

3.8 Plant and animal cells

1 Read the following paragraph and fill in the gaps with words from the box below. Each word may be used once, more than once, or not at all.

All living things are made from .... Each cell uses energy from .... to stay alive. It also uses .... for growth and .... Most plant cells can make their own nutrients using .... Animal cells need to absorb .... from their surroundings.

respiration	photosynthesis	cells	repair	nutrients

2 Label the six parts of the plant cell shown below.

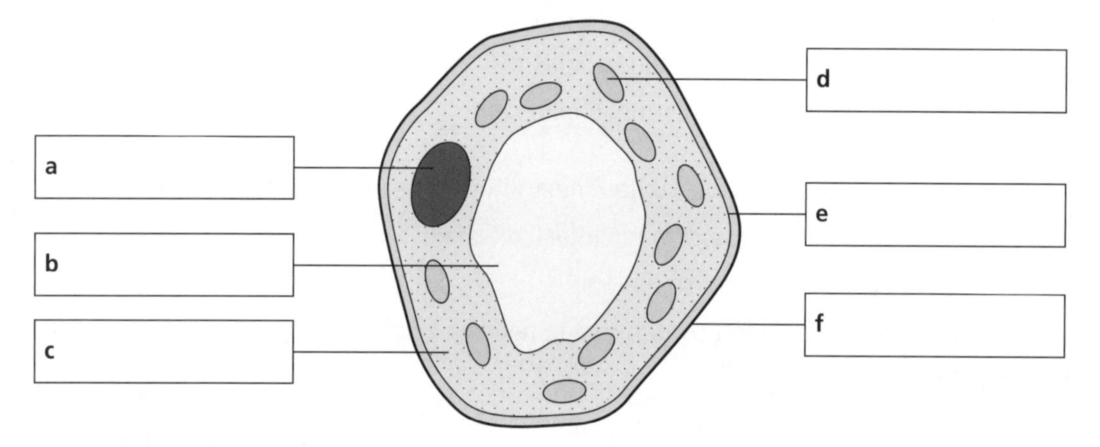

3 Draw lines to match each cell component to its role in the cell.

Cell component
Cell membrane
Chloroplast
Nucleus
Cell wall
Cytoplasm

Role
Absorbs light and allows photosynthesis to take place.
Controls what enters or leaves the cell.
Where most chemical reactions take place.
Controls the activities of the cell.
Stops the cell bursting when its vacuole fills with water.

E

These structures are found in both plant and animal cells. They are mitochondria. Respiration takes place inside them.

Some cells have more mitochondria than others. Which would you expect to contain most mitochondria: muscle cells or skin cells? Explain why.

3.9 Specialised cells

1 Read the following paragraph and fill in the gaps with words from the box below. Each word may be used once, more than once, or not at all.

In plants and animals, each type of cell is .... to suit the .... it does. Cells specialise by having .... shapes and .... and by having different ....

sizes	specialised	different	job	contents

2 Link each specialisation to the correct cell type.

Cell type
Red blood cell
Muscle cell
Fat cell
Bone cell
Root hair cell

Specialisation
Contains fibres which can make themselves shorter.
Contain a large oil droplet which acts as an energy store.
Contains haemoglobin to transport oxygen around the body.
Has a long, thin side branch to absorb water and minerals.
Produces fibres that attract minerals to make a rigid solid.

3 Identify these types of animal cell.

a It is small and flexible so it can squeeze through narrow blood vessels. ....

b It is large and spherical so it is useful for storage. ....

c Cells like this can work together in large groups. They cause movement by making themselves shorter. ....

4 A round cell has a small surface area. Red blood cells and root hair cells both have large surface areas. Explain why each cell needs a large surface area.

....

....

....

....

....

E

Identify each of these types of plant cell.

a As these cells get older they lose most of their components and form hollow tubes which transport water from the roots to the leaves.

b These cells form living tubes to transport sugar to any plant cells which cannot make their own.

c These cells have a very large surface area so they absorb water and minerals efficiently.

3.10 Nerves

1 Read the following paragraph and fill in the gaps with words from the box below.

Nerves are made up of specialised .... They carry .... signals to your brain from .... cells like the sound-detecting cells in .... and the .... cells in your eyes. Your .... converts these signals into sounds and .... you can understand. Nerves also carry .... signals from your brain to your .... to cause ....

light-detecting	cells	electrical	images	chemical	ears	muscles
	brain		movement		sensory	

2 The diagram shows how nerves connect to muscles.

a What type of signal is carried through the nerve cell at X? ....

b What carries a signal between a nerve and muscle cell at Y? ....

3 A bladder is a muscular bag that stores urine. The diagram shows two pig's bladders in separate containers. When a nerve connected to bladder A is stimulated, the bladder contracts. The liquid around bladder A is added to bladder B.

Explain why bladder B contracts.

....

....

....

....

....

....

4 Rahid is having a hearing test. He must raise his hand if he can hear a sound. Put the sentences below in order to describe what makes him raise his hand (label them 1–6).

a Nerves pick up the chemical and send an electrical signal to Rahid's brain.

b The hairs on Rahid's sound-detecting cells vibrate.

c Nerves in Rahid's arm release a chemical which makes muscles contract.

d The sound detecting cells release a chemical.

e Rahid's brain sends signals along nerves to his arm muscles.

f His muscles pull on bones to raise his arm.

3.11 Tissues and organs

1 Read the following paragraph and fill in the gaps with words from the box below. Each word may be used once, more than once, or not at all.

Your body uses organs like your .... to do different jobs. Each organ is made up of .... A tissue is a group of similar .... The cells in each tissue are .... to make them better at doing .... jobs.

tissues	specialised	heart	different	cells

2 Use lines to match each type of tissue to its function.

Tissue
Skin surface cell tissue
Fatty tissue
Bone tissue
Muscle tissue
Blood tissue

Function
Stores energy and reduces heat loss through the skin.
Gives support and structure to limbs.
Keep out micro-organisms.
Brings nutrients and oxygen to other tissues.
Causes movement.

3 Decide whether each statement describes a cell, tissue, organ, or organ system.

a Lungs carry out gas exchange. ....

b Muscle contains cells which can contract. ....

c The mouth, stomach and intestines digest food. ....

d The heart contains several different tissues. ....

e It always contains a nucleus, cell membrane and cytoplasm. ....

4 Your neck connects your head to the rest of your body. List four tissues that you would expect to find there, and explain their role.

....

....

....

....

....

E

Scientists are excited about stem cells. They could be used to grow new tissues and organs for people who need them.

a When your body started to grow it only contained stem cells. What controls the way stem cells divide and differentiate?

b What makes cells build different components and take on different shapes?

4.1 Habitats

1 Write 'desert' or 'rainforest' under each animal to show where it is most likely to be found.

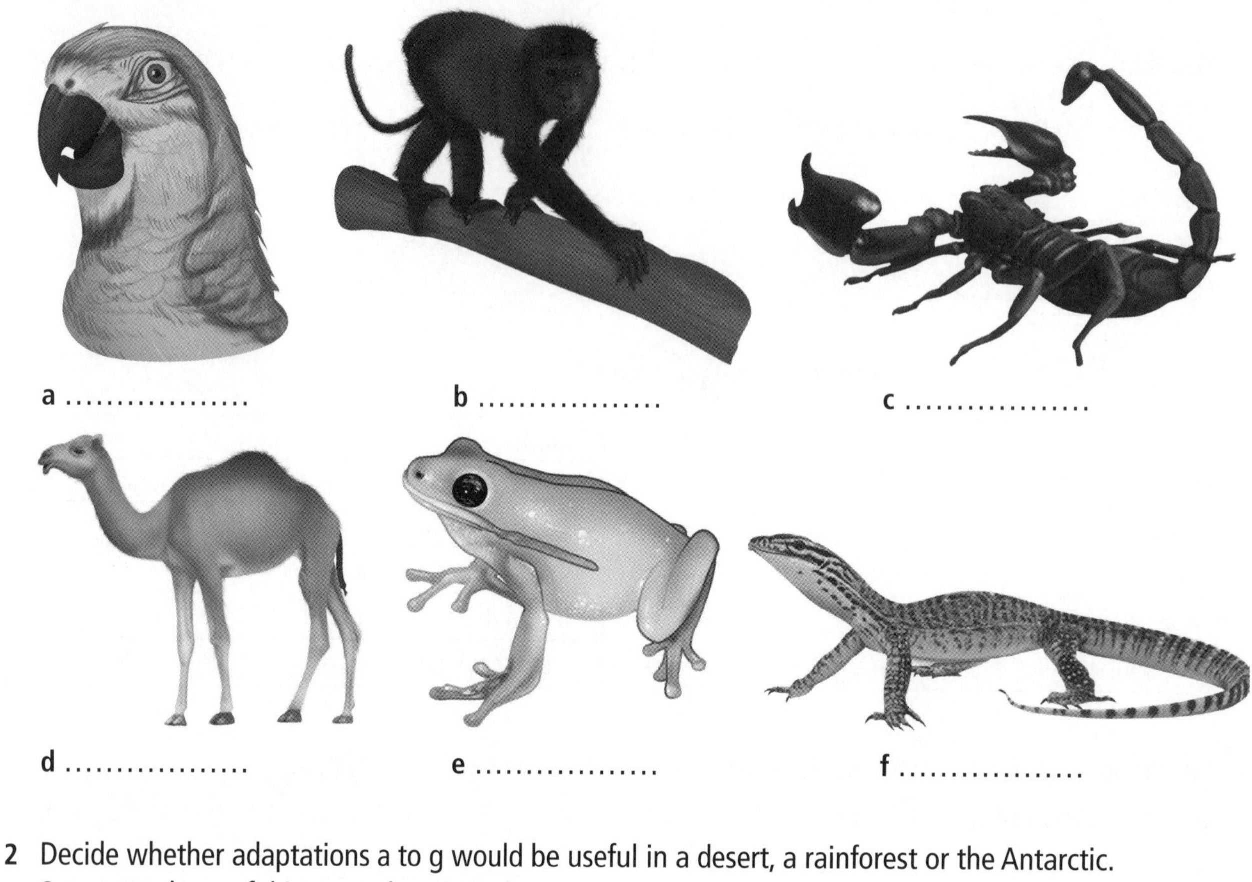

a ....

b ....

c ....

d ....

e ....

f ....

2 Decide whether adaptations a to g would be useful in a desert, a rainforest or the Antarctic. Some may be useful in more than one place.

a A thick layer of fat under the skin ....

b Ability to live in underground tunnels ....

c Ability to climb easily ....

d Large energy stores ....

e Large ears. ....

f Strong arms. ....

g Wide feet ....

3 Write down three adaptations visible in this drawing and explain how they help the polar bear to survive.

....

....

....

....

E

Design a new animal that would be well adapted to life in a desert. Draw a picture of the animal, label its main adaptations and explain why they are useful.

4.2 Food chains

1 Draw lines to match each word to the correct definition.

Word
Predator
Prey
Consumer
Producer
Herbivore
Carnivore
Scavenger

Definition
Eats other living things to obtain nutrients.
Hunts and eats other animals.
Is hunted and eaten by other animals.
Only eats plants.
Makes its own food.
Only eats animals that have already been killed.
Only eats animals.

2 The following food chain is found in many parts of America.
maize → mice → snakes → wild dogs

a Name the producer in this food chain. ....

b Name a consumer which is also a herbivore. ....

c Name a consumer which is both predator and prey. ....

d Explain why all the arrows in a food chain point away from the producer.

....

....

....

e A disease kills the mice. Explain how the maize harvest will be affected.

....

....

....

3 Food chains like the one below are found on many grasslands.
producer → herbivore → carnivore

Fill in the gaps to show what happens if the carnivores are hunted.

The population of predators is reduced so they need less food.

a The herbivore population ....

b The biomass of producers ....

The herbivore population may fall if producers can't regrow fast enough to feed them.

E

Explain how the energy stored in a producer's leaves gets passed along in a food chain such as this one: **acacia tree → giraffe → lion**

4.3 Feeding ourselves

1 Use the words from the box below to complete the following paragraph.

Humans upset natural food .... by destroying ...., causing .... and spreading invasive plants and .... to new habitats. We can reduce this damage by growing more food on less land and reducing our use of ....

pollution	habitats	animals	chains	chemicals

2 Complete the table below to show how each method can increase the mass of food grown.

Method used	How it can increase the mass of food grown	How it can damage natural food chains
Use of fertilisers		
Use of herbicides		
Use of insecticides		

3 Label the diagram below to show why rice and fish both grow faster when they are grown together.

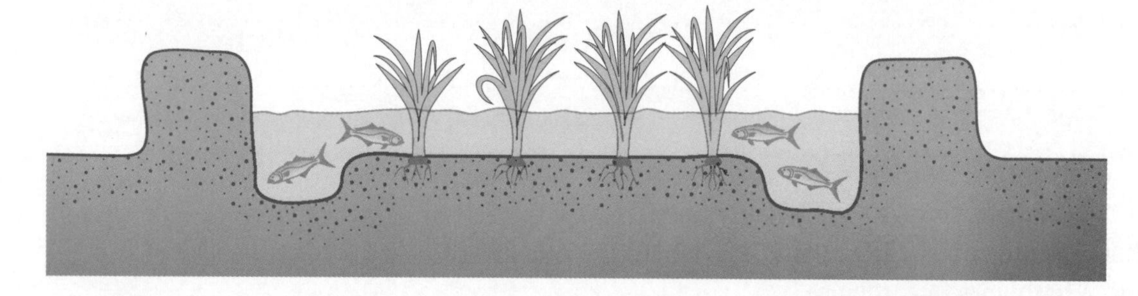

4 Any method that cuts pollution, or reduces the amount of land needed to grow crops, has a positive effect on natural food chains. Write ***P*** after the statements that are positive and ***N*** after those that are negative.

a The human population is growing rapidly so we need to produce more food.
b Micro-organisms can be grown in tanks to reduce the land used to grow crops.
c Insects can be spread around the world when crops are exported.
d Plants can be grown in high-rise greenhouses to save land.
e Fish and rice can be produced together to improve their growth.
f Crops can be harvested from forests to avoid destroying them.

E

Explain why world food shortages could be reduced by growing more algae.

4.4 Changing the planet

1 Use the words from the box below to complete the following paragraph.

The mixture of gases that surrounds the Earth is called the .... We rely on it for our survival. Three .... have damaged the atmosphere. Chemicals called CFCs reduced the amount of .... in the upper atmosphere and allowed more harmful .... light to reach the Earth. Acidic gases from burning .... have reduced the pH of .... and carbon dioxide has caused global ....

fuels	warming	ultraviolet	atmosphere	pollutants	rainwater	ozone

2 The bar chart shows how the area of the ozone hole over Antarctica has changed.

a Calculate the increase in its size between 1980 and 2000.

....

b Describe what has happened to the hole since the year 2000.

....

c Suggest a reason for this change.

....

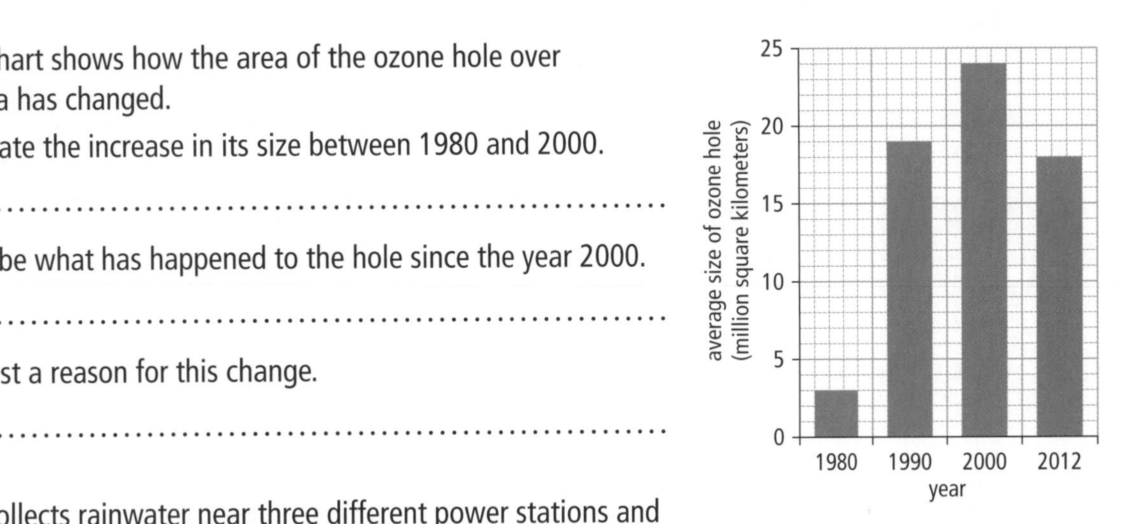

3 Paavan collects rainwater near three different power stations and measures its pH. Rain is acidic if the pH is less than 7.

Power station fuel	pH of rainwater
Coal	4.5
Coal with pollution controls	6.5
Gas	6.5

a Name two gases released when coal and gas burn.

....

b Describe two ways of reducing acid rain.

....

....

....

E

The percentage of carbon dioxide in the atmosphere is increasing.

a Explain how this will raise the temperature of the atmosphere.

b Describe two ways this global warming could be reversed.

4.5 Preventing extinction

1 Draw lines to match each word to the correct definition.

Word
Conservation
Sanctuary
Captive breeding

Definition
A place where species can be protected
Breeding animals in zoos
Preventing extinction

2 Suggest the best conservation method for each of these animals.

a A bird whose numbers dropped when rats invaded its island. ....

b An animal with valuable fur that is hunted by poachers. ....

c An animal threatened by deforestation. ....

d An animal that only survives in zoos. ....

3 Jaguars used to be common in South American forests. Many of these forests have been cleared to make room for farms. The remaining jaguars are often hunted because their fur is very valuable.

a List two reasons why jaguar numbers have fallen in recent years.

....

....

b Suggest how the species could be conserved.

....

....

4 The Galapagos Islands contain many species that are not found anywhere else on Earth, such as the giant tortoise. In the past, goats were brought to the islands by sailors. The goats ate the same plants as the giant tortoise.

a Suggest how the goats affected the tortoise population.

....

b Tortoise numbers are growing on one island. Suggest why.

....

....

E

Leatherback turtles travel thousands of miles before they return to the beaches where they were born to mate. They feed mainly on jellyfish.

Do some research. Find two reasons why these animals are endangered and one way they can be protected.

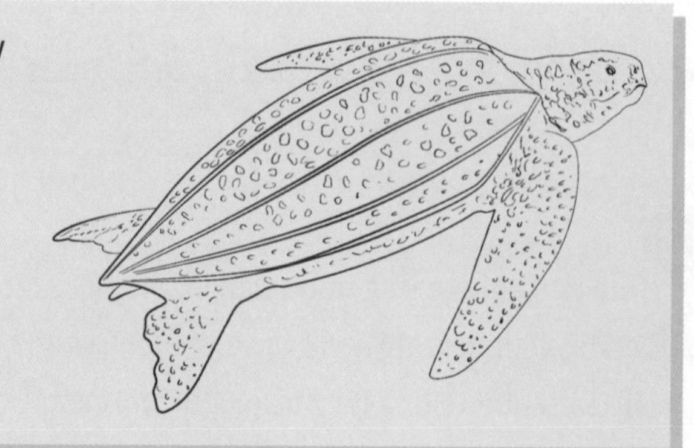

4.6 Obtaining energy

1 Draw lines to match each energy term to the correct definition.

Energy term
Biofuel
Geothermal energy
Solar cells
Renewable energy

Definition
Devices that use light to produce electricity
Energy sources that are constantly replaced
A fuel made by living things
Heat from hot rocks under the Earth's surface

2 Choose the correct word to replace the incorrect ones used in the following sentences.

Fossil fuels are non-renewable because they cannot be ~~reused~~ ....

Renewable resources are ~~never~~ .... replaced.

3 Read statements a-g. Write ***T*** next to the true statements and ***F*** next to the false ones. Then re-write any false statements correctly.

a Biofuels will soon run out. ....

b Solar cells produce a lot of electricity. ....

c Oil, gas and cow dung are all examples of fossil fuels. ....

d Energy from the Sun makes winds blow. ....

e Plants can be planted to produce biofuels as quickly as they are burnt. ....

f Geothermal energy could be used anywhere in the world. ....

g Fossil fuels were useful sources of energy in the past. ....

....

....

....

....

....

4 Choose the best renewable energy source for places a to e.

a An isolated village in the desert. ....

b A town in East Africa where there are hot rocks close to the surface. ....

c A maize growing area that has lots of waste plant material. ....

d A village near the coast where it is always windy. ....

e A mountain village where it rains a lot. ....

f A large animal farm in Europe. ....

E

The amounts of fossil fuels used for electricity generation is falling. Explain which renewable energy sources could be used to replace them in your country.

4.7 Growing fuels

1 Choose the best words to complete the following sentences.

Biofuels are .... because they are made by living things. The two main biofuels are biodiesel, which is made from ...., and bioethanol, which is made from ....

2 Label the diagram to show why biofuels could be carbon-neutral.

3 When palm oil is used to make biofuels the process is not carbon-neutral. Describe two ways extra carbon dioxide is released.

....

....

....

....

4 Waste plant materials can be broken down to form sugars. Then these sugars can be turned into bioethanol.

a Name a micro-organism that converts sugar to ethanol. ....

b Which type of decomposer converts plant cell walls to sugar. ....

c Name the useful chemicals these decomposers produce. ....

5 Complete the table to compare the advantages and disadvantages of two biofuels.

Fuel	Advantages	Disadvantages
Biodiesel from algal oil		
Bioethanol from sugar cane		

6 Biodiesel can be made from palm oil or algal oil. At the moment, most biodiesel is made from palm oil. Write a paragraph to convince people to make it from algal oil instead.

....

....

....

....

5.1 Variation

1 Choose words from the box below to complete the following sentences.

Some of our features make us different from everyone else. These ....

differences can be used to .... people. Features like our body sizes can take

any value within a certain .... They show .... variation.

Other features fall into a few distinct categories so they show .... variation.

range	discontinuous	unique	identify	continuous

2 Write ***C*** after the features that show continuous variation and ***D*** after the features that show discontinuous variation.

a Blood group ....

b Height ....

c Foot length ....

d Shoe size ....

e Body mass ....

3 The table shows the distribution of blood groups in India. On a piece a graph paper, use the data to plot a frequency chart.

Blood group	Number of people (%)
O	39
A	23
B	32
AB	6

4 The table shows the distribution of test scores in a class. Use the data to plot a frequency chart on graph paper.

Score (%)	Tally	Frequency
0–20	1	1
21–40	~~1111~~	5
41–60	~~1111~~ 1111	9
61–80	1111	4
81–100	1	1

E

Passports often get stolen so that other people can use them. Many people have similar faces, so it can be hard to identify people from their photographs.

a Explain why biometric data prevents the use of stolen passports.

b Suggest what biometric data would be best for use at airports.

5.2 Causes of variation

1 Variation can be caused by the genes you inherit, the environment you experience or a combination of the two. Decide which features belong to each category and write their letters into the correct section of the Venn diagram.

a Blood group
b Height
c Behaviour
d First language
e Body mass
f Eye colour
g Test scores
h Risk of disease

inherited variation

environmental variation

2 Read statements a–g. Write ***T*** after the ones that are true and ***F*** after the ones that are false. Write the correct version of false statements on the lines below.

a Your genes control the activities of your cells.
b We inherit genes from both parents so we inherit some of their features.
c Twins always have identical genes.
d Identical twins develop differences when their environments differ.
e A poor environment can prevent your genes from keeping you healthy.
f Every cell in your body contains identical genes.
g Cells specialise by switching on all their genes.

....

....

3 Explain why the average height of students is increasing in many parts of the world.

....

....

....

4 Scientists want to know whether your genes or your environment has the biggest influence on your behaviour. To answer this question they compare identical twins that were separated at birth with identical twins who grew up together.

a Suggest why identical twins separated at birth are ideal for these studies.
b What results would you expect if inherited variation turned out to be more important than environmental variation in determining your behaviour?

....

....

....

5.3 Species

1 Choose the best words from the box below to complete the following sentences.

A species is a group of plants or animals that share many of their .... Members of the same .... can breed with each other and produce offspring that are also able to breed. Each species has been given a two-part .... name that is used all over the .... Members of different species do not usually breed with each other. Those that do usually have .... offspring called ....

infertile	world	Latin	hybrids	characteristics	species

2 The two-part Latin name for a zebra is *Equus zebra*. Decide whether the animals below are from the same species, similar species or very different species from the zebra. Write their names in the correct columns in the table.

Equus ferus	*Tetracerus quadricornis*	*Syncerus caffer*	*Equus africanus*

Same species	**Similar species**	**Very different species**

3 Read statements a-f. Write ***T*** after the ones that are true and ***F*** after the ones that are false. Write the correct version of false statements on the lines below.

a Variation between species is usually greater than variation within species.

b Similar species share the same two-part Latin name.

c Members of the same species should produce fertile offspring.

d Members of the same species always look similar.

e A hybrid is born when animals from different species breed.

f Members of different species usually have fertile offspring if they breed.

....

....

....

....

E

A dzo is a cross between a yak and domestic cow. It is much stronger than either of its parents and grows faster.

a Suggest why dzos have not been bred to form large herds, despite being more useful than both cows and yaks.

b Yaks and dzos can look quite similar. Suggest how scientists could distinguish between them.

5.4 Classification

1 Choose the best words from the box below to complete the following sentences.

Scientists use similarities and .... to put living things into .... This is .... The major groups of animals are vertebrates, which have .... and invertebrates which do not. Most animals are .... These are subdivided into smaller and smaller sub-groups with different .... The smallest sub-group is a ....

groups	characteristics	differences	classification	species	backbones	invertebrates

2 The diagram shows one animal from each invertebrate group.

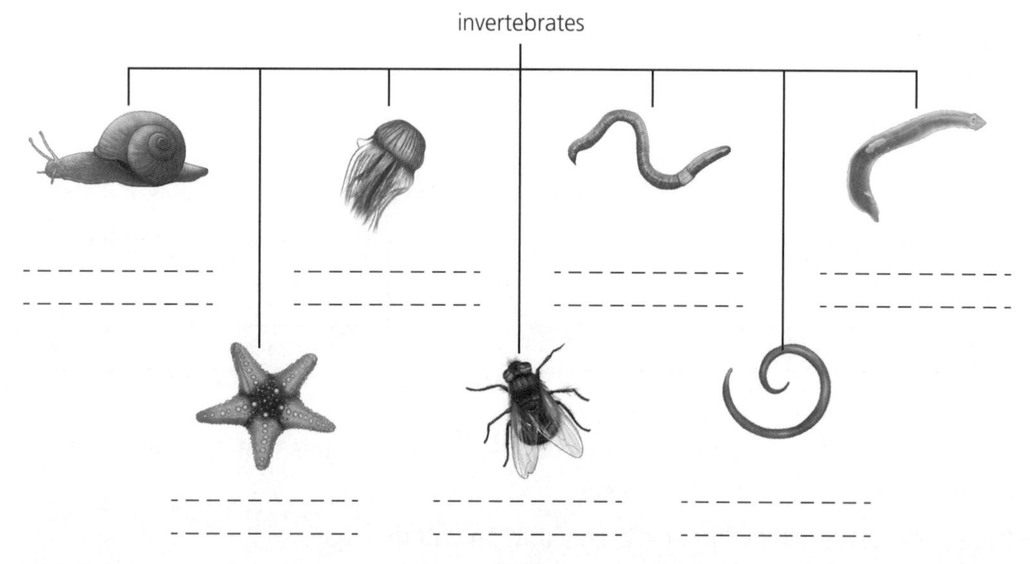

Write the name of the group and the main characteristic that animals in that group share.

3 Most invertebrates are arthropods. The diagram below shows four groups of arthropods.

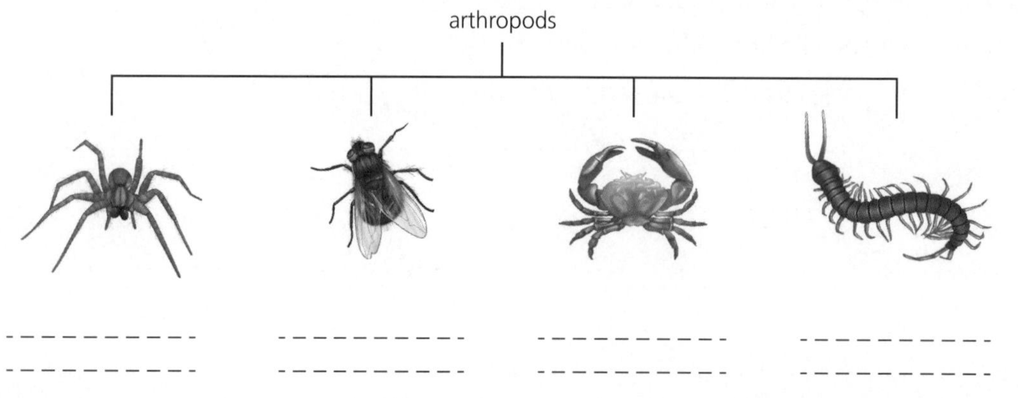

Write the name of the group and the main characteristic that animals in that group share.

E

The drawing below shows a type of invertebrate.

Write what type of invertebrate it is and explain how you can distinguish it from other invertebrate groups.

5.5 Vertebrates

1 The diagram shows one animal from each vertebrate group.

In the space below each drawing, write the name of the group and two important characteristics that the animals in that group share.

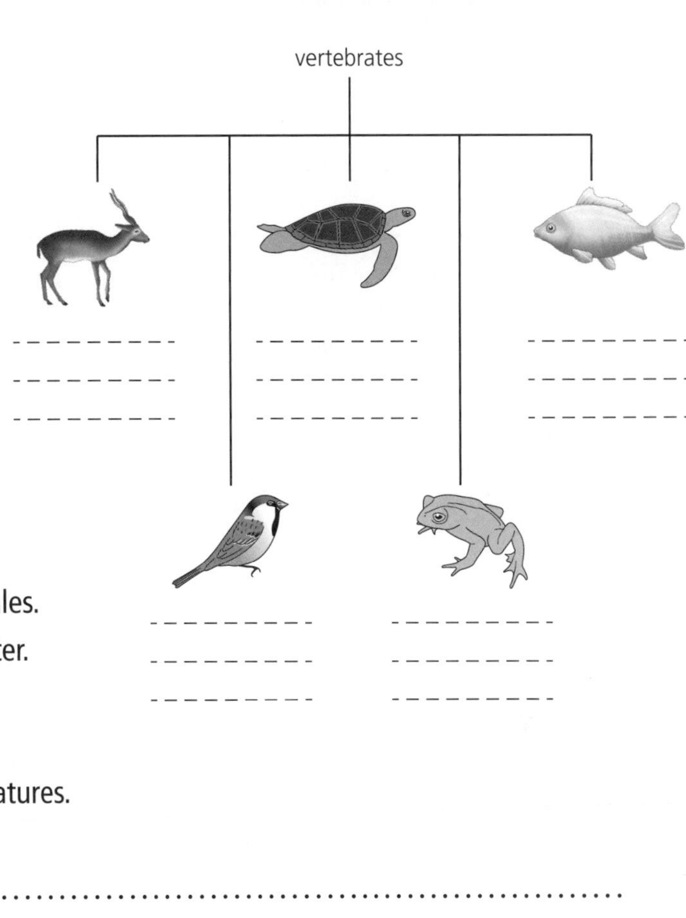

2 Read statements a–g. Write ***T*** after the ones that are true and ***F*** after the ones that are false. Write the correct version of false statements on the lines below.

a Vertebrates can be classified using their body coverings, the way they reproduce and the way they take in oxygen.

b Snakes are reptiles because they have hard scales.

c Reptiles can only reproduce where there is water.

d Birds and mammals keep their bodies warm.

e Fish are the only vertebrates which have gills.

f Some animals do not share all their group's features.

g Whales are fish but feed their young on milk.

....

....

....

3 The Echidna is a small vertebrate covered with hair and spines. Three weeks after mating, females produce an egg. This is kept in a pouch. After 10 days the egg hatches and the baby echidna feeds on its mother's milk.

a Which vertebrate group do echidna belong to? ....

b Describe two features this vertebrate group shares with echidnas.

....

....

c Describe one feature that makes echidnas different from the rest of the group.

....

E

The drawing shows an artist's reconstruction of an extinct vertebrate called archeopteryx. It had claws, feathered wings, a long bony tail and a beak full of teeth.

a Give one reason why archaeopteryx is difficult to classify.

b Which two vertebrate groups does archaeopteryx share features with?

5.6 Classification of plants

1 Complete this table to show the features the plants in each group share.

Type of plant	moss	ferns	conifers	flowering plants
Roots and veins?				
Spores or seeds?				
Cones, flowers or neither?				

2 Use the table to complete this key.

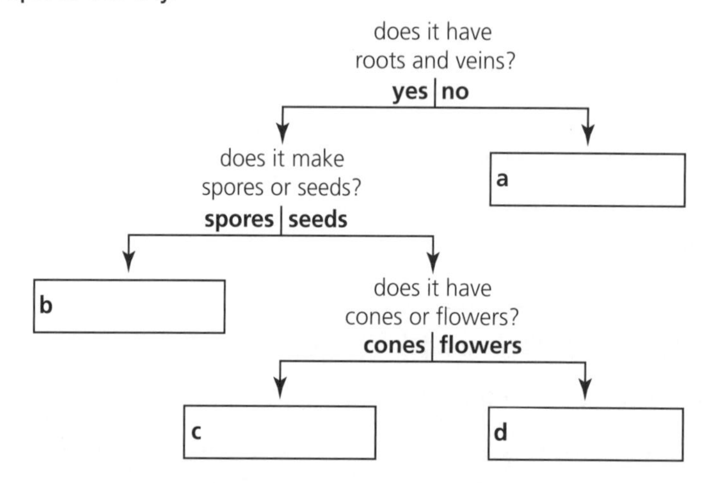

3 Identify the groups plants A to D belong to.

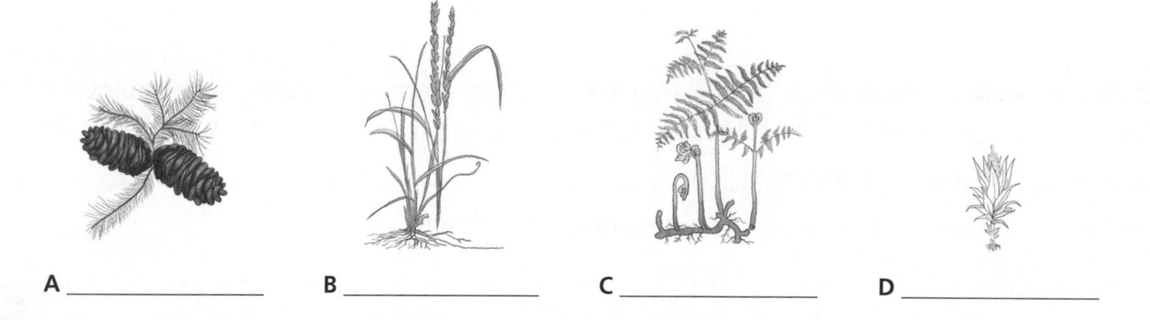

A ______________ B ______________ C ______________ D ______________

E

Algae are important in the oceans' ecosystems. They produce nutrients to support food chains and add oxygen to the water. The largest algae are called seaweeds. Unlike plants, they cannot survive on land.

Describe two differences between flowering plants and seaweeds that make flowering plants better adapted to life on land.

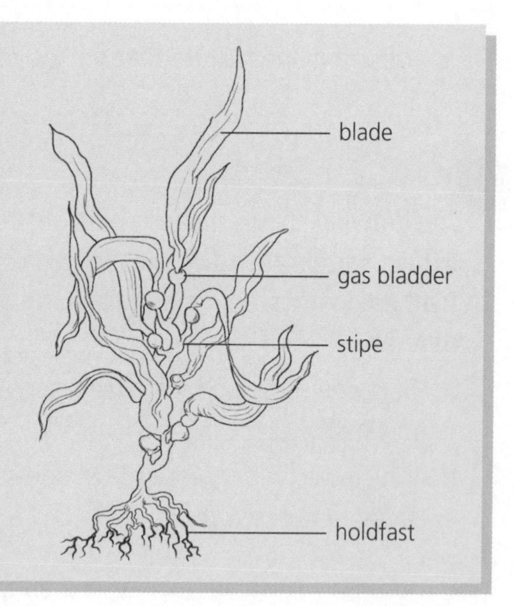

6.1 Why we need plants

1 Read the following paragraph and fill in the gaps with words from the text box below. Each word may be used once, more than once, or not at all.

The material living things are made of is called .... Plants build new biomass using .... During photosynthesis, light .... is absorbed. It makes .... from the air react with .... from the soil. The products are glucose and .... Plant cells can release energy from glucose molecules, using ...., or use them for growth.

photosynthesis oxygen energy respiration carbon dioxide biomass water

2 Decide whether the following statements refer to photosynthesis, respiration or both. Write each letter in the correct part of the Venn diagram.

a Uses carbon dioxide from the atmosphere.

b Can be used to produce starch.

c Takes place in all living cells.

d Releases energy.

e Releases oxygen.

f Stores energy.

g Uses oxygen from the atmosphere.

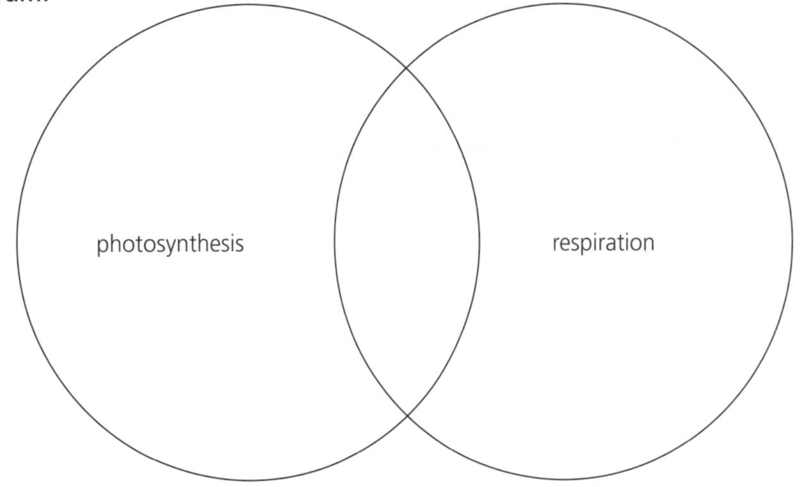

3 Draw lines to match each part of a leaf to its role in photosynthesis.

Part of leaf
Palisade cells
Stomata
Xylem vessels
Mesophyll cells

Role in photosynthesis
Hollow tubes that carry water up from the roots
Contain most chloroplasts to absorb light
Form a spongy layer which gases can diffuse through
Tiny pores which let gases in and out of leaves

E

Fiona lives in Europe, where the summers are warm and sunny and the winters are cold and cloudy.
A geranium plant grows on the windowsill in her classroom. She tests one of its leaves for starch in summer and repeats the test in winter.
Explain why the results are different in winter and summer.

Season	Starch test results
Summer	Leaf turns blue-black
Winter	Leaf turns dark brown

6.2 Asking scientific questions

1 Read the following paragraph and fill in the gaps with words from the box below. Each word may be used once, more than once, or not at all.

Scientific questions involve two .... , one that can be .... to answer the question, and one that can be .... to see what effect that change has. There is usually .... than one variable that can be measured. It is important to .... the measurements to check that the results are ....

more	repeat	changed	reliable	variables	measured

2 Read the following statements. Write ***S*** next to the scientific questions and NS next to the ones which are not scientific.

- a Do plants with longer roots grow faster?
- b Are tall plants better for farmers?
- c Is it worse for plants to have fewer leaves?
- d Does a plant gain mass faster in brighter light?
- e Can a thicker stem support more mass?

3 Raman wants to find out whether the volume of water a plant receives affects its growth.

- a What variable should he change? ....
- b What could he measure? ....
- c List two variables he should keep the same. ....

....

4 Write ***B*** next to the results that should be plotted on a bar chart and ***L*** after the ones that should be displayed as a line graph.

- a The number of petals on different types of flower.
- b The heights of several plant species found in one field.
- c The height gain of plants grown at different temperatures.
- d The mass of fruit produced by plants with different numbers of leaves.
- e The growth rates of plants with different numbers of stomata in their leaves.

E

Sai wants to find out how the amount of carbon dioxide available affects the rate of photosynthesis. She has beakers of water containing different amounts of carbon dioxide, pondweed, beakers, measuring cylinders and funnels.

- a Describe how she should set up her apparatus.
- b Predict what results she would obtain and give reasons for your answer.

6.3 Water and minerals

1 Read the following paragraph and fill in the gaps with words from the box below. Each word may be used once, more than once, or not at all.

Plants need a continual supply of .... They use it for .... and it evaporates from their leaves. If plants lose more water than they can .... from the soil, their cell .... shrink. This makes each .... flaccid. Flaccid cells cannot .... themselves so they make the plant ....

vacuoles	support	cell	wilt	water	absorb	photosynthesis

2 Label the arrows on the diagram below to explain how water moves through a plant.

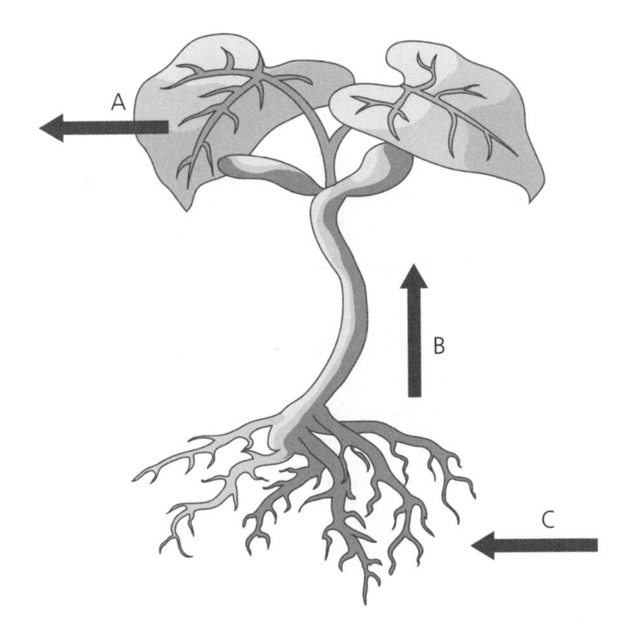

3 Draw lines to join each observation about transpiration to the correct explanation.

Observation
Plants continually lose water
There are pores on the undersides of leaves
Root hairs give plants a large surface area
Water flows from the roots to leaves
Plants lose less water at night

Explanation
to allow carbon dioxide to enter them
to help them absorb more water
because their stomata close
because it evaporates from their leaves
because it is pulled up xylem tubes to replace the water that evaporates

E

Many house plants die if they over-watered. The excess water fills the air gaps between soil particles and reduces the amount of oxygen roots can take in.

a Explain why root cells need oxygen to stay alive.

b How would a shortage of oxygen affect mineral uptake?

7.1 Food

1 Draw lines to match each nutrient to the correct role.

Nutrient
Carbohydrates
Fats
Proteins
Vitamins
Minerals

Role
Used to build cell membranes and a good source of energy
Help chemical reactions take place in your cells
Most of our energy intake should come from these
Help cells to function properly and strengthen bones and teeth
Essential for growth and repairing cells

2 Write the names of the main nutrients each of these foods contain. Choose up to two nutrients from carbohydrates, proteins and fats.

a Sugar ....

b Oil ....

c Chocolate ....

d Meat ....

e Cheese ....

f Banana ....

g Nuts ....

h Rice ....

3 Rashid thinks that fat is bad for you. Give three reasons why we need fat in our diet.

....

....

....

....

4 Proteins, fats and starch are large molecules made by joining smaller molecules together. Complete the table to show what is different about each of these small molecules.

Nutrient	Small molecules joined to make them
Starch	
Proteins	
Fats	

E

In many parts of the world, children eat mainly rice and vegetables.
Which nutrient do they lack and what problems might this cause?

7.2 Managing variables

1 Eniola is testing foods to see which one contains most energy. She uses the apparatus shown on the right.

During the investigation, variables need to be 'changed', 'measured', 'controlled' or 'calculated'. Write the correct choice next to variables a–g.

a Type of food ....

b Mass of food ....

c Volume of water ....

d Distance between food and test tube ....

e Water temperature before food burns ....

f Water temperature after food burns ....

g Temperature rise ....

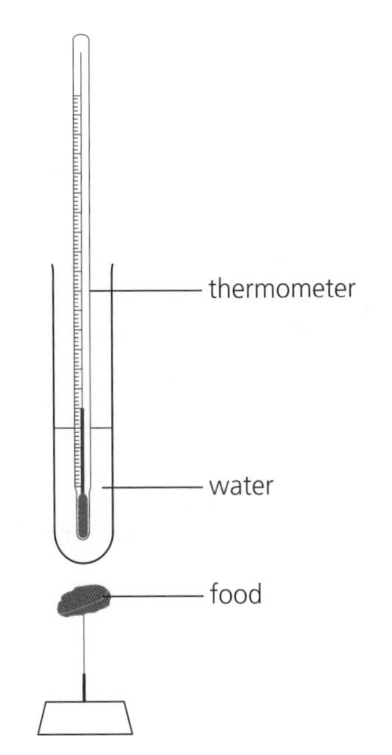

2 Eniola uses three measuring instruments during the investigation. Complete the table to show what each instrument measures and the units.

Measuring instrument	Variable measured	Units used
thermometer		
measuring cylinder		
electronic balance		

3 When Eniola burned twice as much bread, the temperature rise of the water doubled. Give two reasons why Eniola should only burn small quantities of food.

....

....

....

....

E

Eniola compared the energy content of bread, chicken breast and cheese.

Food tested	Mass of food (grams)	Temperature before (°C)	Temperature after (°C)	Temperature rise (°C)	Temperature rise per gram (°C per gram)
bread	1.5	21	51		
chicken	1	22	68		
cheese	2	22	88		

Complete the table of results and suggest a reason for the differences between the foods.

7.3 A balanced diet

1 Read the following paragraph and fill in the gaps with words from the box below. Each word may be used once, more than once, or not at all.

A balanced diet contains every essential ...., in the correct .... These include ...., essential ...., vitamins and .... You also need starchy carbohydrates like .... to supply ....

fatty acids	nutrient	energy	minerals	rice	proportions	proteins

2 The table shows the fat contents of three types of nut.

Type of nut	Total fat (g /100 g)	Saturated fat (g /100 g)	Unsaturated fat (g /100 g)
brazil	63	16	
coconut	36	29	
almond	40	5	

a Calculate how much unsaturated fat each type of nut contains and write it in the table.

b Which nut would you recommend for a patient at risk of heart disease?

....

3 The table shows the nutrients in bars of milk chocolate.

Nutrient	Mass present (g /100 g)
sugar	57
protein	8
fat (mostly saturated)	30

Give two reasons why chocolate should not be your main source of carbohydrate.

....

....

....

4 Dev does not eat meat. He gets most of his protein from beans. Explain why he needs to add other sources of protein to his diet.

....

....

E

Assume that you need 8000 kJ per day and only 25% should come from fat.
Fat releases 37 kJ/gram. Calculate how many grams of fat you should eat each day.

7.4 Deficiencies

1 Read the following paragraph and fill in the gaps with words from the box below. Each word may be used once, more than once, or not at all.

In the past poor diets made many people .... Scientists used creative thinking to .... why this happened. They suggested that illnesses like .... were .... diseases. They were caused by a .... of essential .... or minerals. When the missing .... were supplied, they recovered.

scurvy	nutrients	explain	lack	ill	deficiency	vitamins

2 Draw lines to match each deficiency disease to the missing nutrient.

Deficiency disease
Anaemia
Kwashiorkor
Scurvy
Beri-beri
Rickets
Night blindness

Missing nutrient
Protein
Vitamin C
Iron
Vitamin A
Vitamin B1
Vitamin D

3 Read what each patient says and decide which deficiency disease they have.

a I feel tired all the time and my friends say I look quite pale. ....

b I can't see in the evenings but my vision is fine in sunlight. ....

c My gums are bleeding and my teeth have started to fall out. ....

d I get confused and find it hard to move my legs. ....

e I've broken my arm and my legs bend outwards at the knee. ....

f My legs are very thin but my tummy and feet are swollen. ....

E

Hassina does not feel well. Her doctor tests her blood. The results are shown below.

Vitamin / mineral	Normal blood (units)	Hassina's blood (units)
Calcium	2.2-2.6	2.4
Vitamin C	0.4-1.5	0.2
Iron	50.0-170.5	30.7
Vitamin A	30.0-65.0	55.4

a List any deficiency diseases Hassina may have.

b What symptoms would you expect Hassina to have?

7.5 Choosing foods

1 Read the following paragraph and fill in the gaps with words from the box below.

In most countries average .... masses are rising and more people are becoming .... Their diets supply more .... than they use and the excess is stored as .... This can cause long-term health problems like ...., cancer, high .... pressure, and .... disease.

heart	blood	fat	diabetes	body	energy	obese

2 Health workers want to improve peoples' diets. Match each action they propose with the correct reason.

Action
Make fatty and sugary foods more expensive
Encourage people to exercise more
Add extra nutrients to common foods
Add extra genes to common crop plants

Reason
so sugars and fats won't make them obese.
to reduce demand for them.
so extra nutrients aren't needed.
so people automatically get a balanced diet.

3 The table shows information about two types of food on sale in supermarkets.

Food	Chocolate biscuits	Fresh berries
Price (US$/kg)	2	6
Special offers / year	5	2
Sugar (grams/kg)	290	100
Saturated fat (grams/kg)	130	0

a Write two reasons why people buy more chocolate biscuits than fresh berries.

....

....

b Suggest two reasons why eating a lot of the biscuits could cause health problems.

....

....

....

4 Many people suffer from deficiency diseases because they can't afford to eat a balanced diet. Describe two ways of increasing the amounts of vitamins and minerals they consume without changing their diets.

....

....

....

8.1 The digestive system

1 Add labels to this diagram of the digestive system.

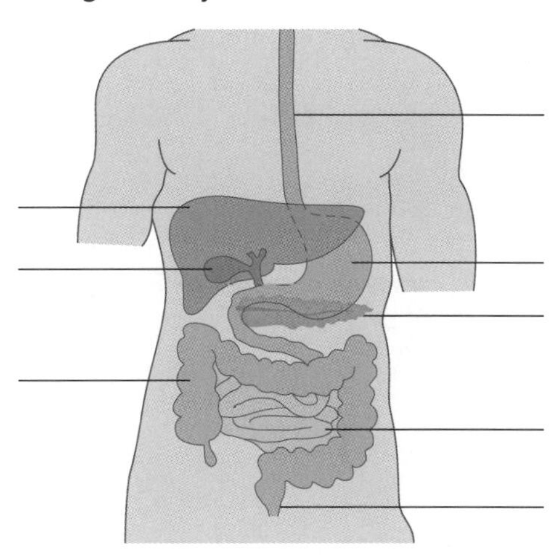

2 Read the following paragraph and fill in the gaps with words from the box below.

Digestion breaks down food into small .... that can enter your .... It happens in your gut (...................... ....). Teeth begin the process. They crush food into smaller .... Then .... help the large molecules in food to .... down. These enzymes are produced in your mouth, stomach, ..., and .... intestine. The small intestine .... digested food and the large .... removes .... from the waste products left behind.

water	alimentary canal	blood	intestine	molecules
small	enzymes	pancreas	break	pieces	absorbs

3 Decide whether statements a–f refer to **mechanical digestion**, **chemical digestion** or **both**.

- **a** Digestion begins in your mouth. ....
- **b** Teeth chew food to break up large pieces. ....
- **c** Chewing increases the surface area of solid foods. ....
- **d** Enzymes help large molecules to break down. ....
- **e** Saliva contains an enzyme which breaks down starch. ....
- **f** Digestion ends in your small intestine. ....

4 We cannot digest fibre. Explain why we still need to eat it.

....

....

E Make a flow chart to show what happens to the starch and fibre in a banana in each part of your alimentary canal.

1 The diagram shows what happens when food is digested.

a On the diagram, label the three large molecules found in food and the small molecules formed when they break down.

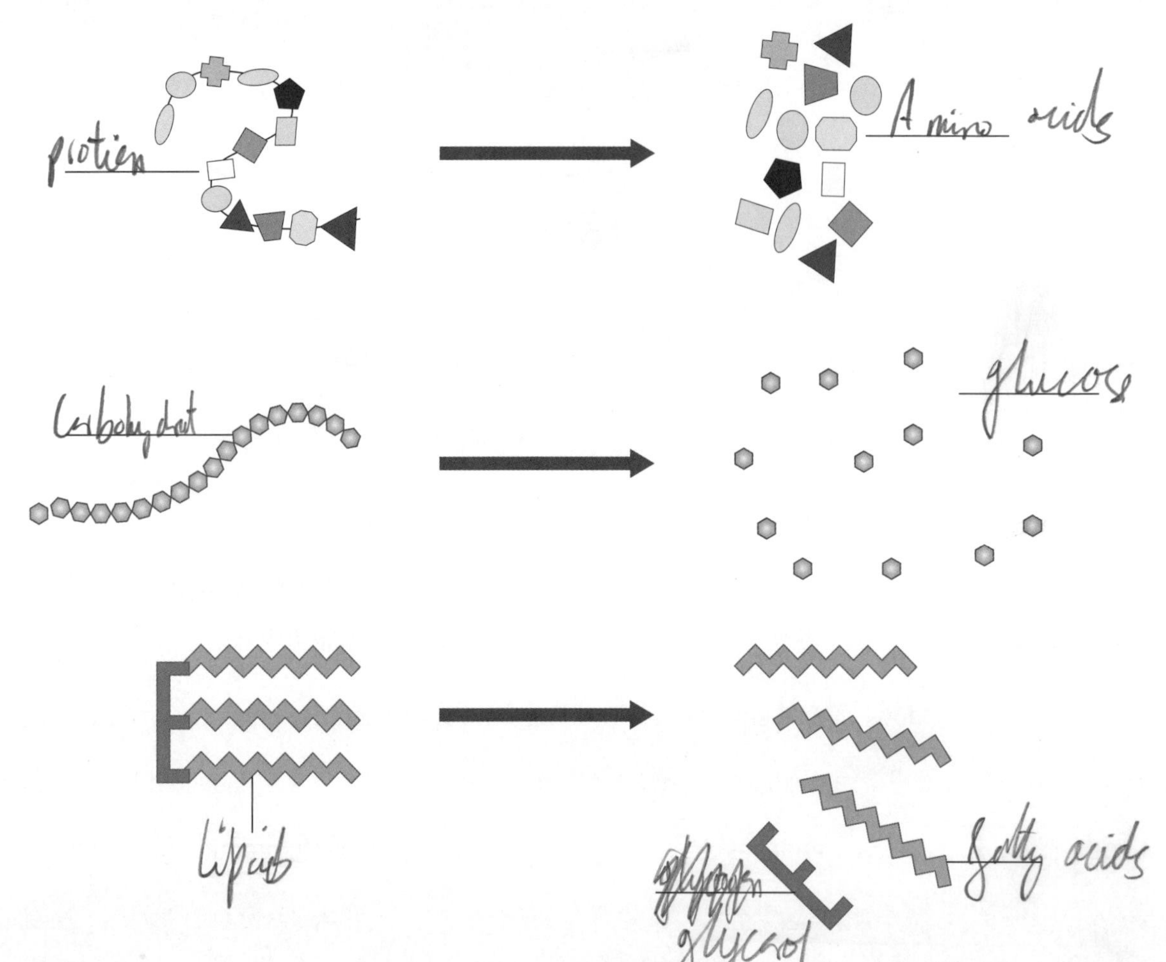

b Add labels to each arrow to show the type of the enzyme involved and where these food molecules are digested.

2 Draw lines to match each word to the correct definition.

Word
Enzyme
Carbohydrase
Emulsify
Bile

Definition
Substance that emulsifies fats to increase their surface area
To break fats into smaller droplets which can mix with water
Biological catalyst used to speed up reactions
Enzyme like amylase which breaks down carbohydrates

3 Solid foods are broken down in the mouth. Explain how this aids digestion.

....

....

E

Petros thinks that amylase will not work as well at high temperatures.
Describe how he could collect evidence to support his idea.

8.3 Using enzymes

1 Draw lines to match each enzyme to one or more uses.

Enzyme
Lipase
Carbohydrase
Protease

Use in the food industry
Breaks down corn starch to make it sweeter
Removes fats from meat or fish
Improves the flavour and texture of fatty foods
Turns milk into a solid curd during cheese-making

2 Write ***T*** after the true statements and ***F*** after those that are false. Then write corrected versions of the statements that are false.

a Enzymes from living organisms can be used to speed up reactions in industry. T

b The active site of an enzyme is the same shape in all enzymes. F

c Enzymes need to be replaced when they finish catalysing a reaction. F

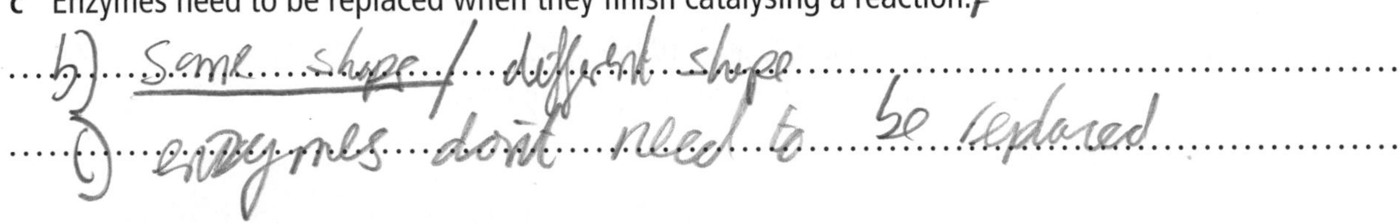

b) Same shape / different shape
c) enzymes don't need to be replaced

3 The diagram shows an enzyme helping to break down sucrose.

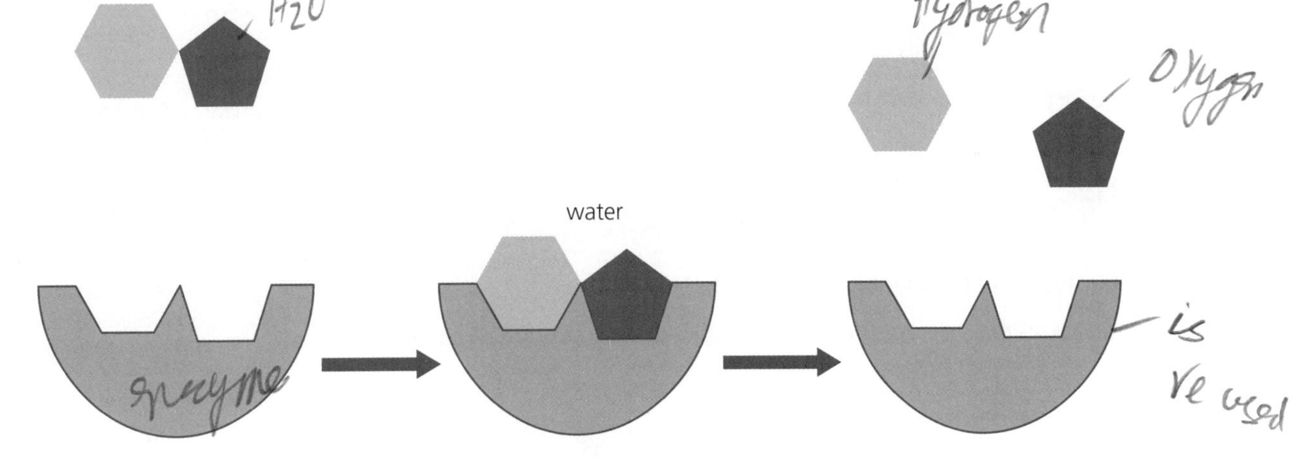

a Label the diagram to show how the enzyme speeds up the reaction.

b Explain why enzymes are only needed in small amounts.

they can be re used over and over unless they get denatured

4 All the enzymes in your digestive system help large molecules to react with water. Use diagrams to explain why more than one sort of enzyme is needed.

1 The diagram shows part of a blood smear.

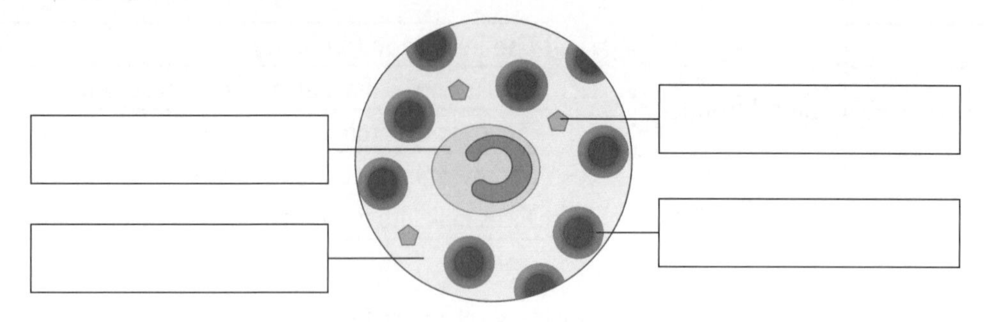

a Label the four main components of blood.

b Explain why most cells in a blood smear look paler in the centre.

....

2 Statements a-f refer to the four main components of blood.
Write the name of the correct component next to each statement.

a Carries oxygen around the body ....
b Needed to fight infections ....
c Packed full of haemoglobin ....
d The largest cells in your blood ....
e A pale yellow liquid ....
f Carries dissolved substances around your body ....
g Helps to form a clot when a blood vessel is damaged ....
h Has a biconcave shape to increase its surface area ....

3 Capillaries carry blood past cells all over your body.

a Label the diagram to show what happens as blood passes cells.

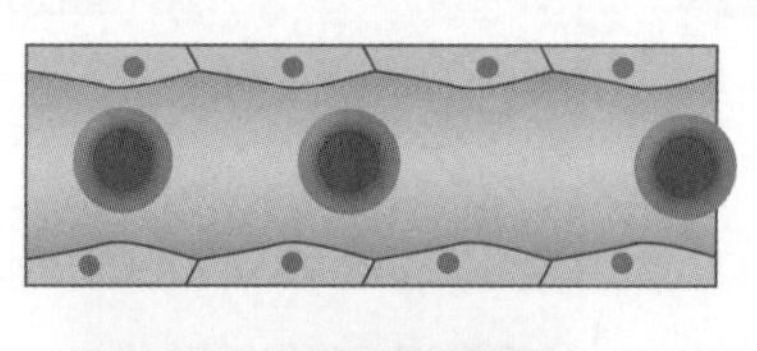

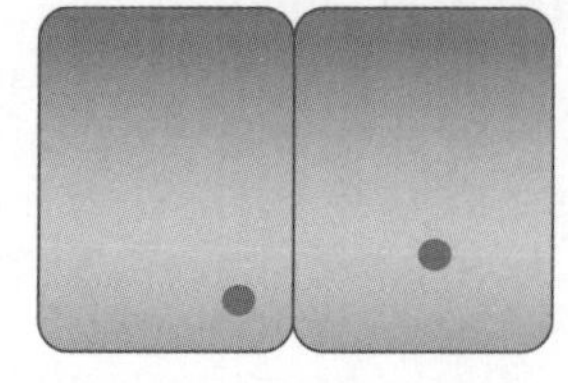

b Explain what makes molecules move in and out of blood.

....

....

E Use a diagram to explain the symptoms of sickle cell anaemia.

9.2 Anaemia

1 Write ***T*** after the true statements and ***F*** after those that are false. Then write corrected versions of the statements that are false.

a Anaemia can make you feel tired all the time.

b Anaemia prevents blood from carrying enough carbon dioxide.

c Anaemic blood contains fewer white blood cells than normal.

d A patient with anaemia has less haemoglobin than normal.

e A high packed cell volume shows that a patient has anaemia.

f Anaemia can be caused by a lack of iron in the diet.

g You can reduce the symptoms of anaemia by eating less red meat.

....

....

....

....

2 The table below shows the results of some blood tests.

Patient	White blood cell count (billions/dm³)	Red blood cell count (billions/dm³)	Haemoglobin level (g/dm³)	Packed cell volume (%)
A	0.007	4.6		47
B	0.002	3.9	114	43
C	0.007	2.6	89	28

a Patient A's blood contains 14.5 grams of haemoglobin per 100 cm^3.
Calculate their haemoglobin level in g/dm^3 (1 $dm^3 = 1000\,cm^3$). Write it in the table.

....

....

b Deduce which patient has anaemia and give two pieces of evidence to support your answer.

....

....

c Only one of the patients is male. Deduce which one and give your reasons.

....

....

3 Modern hospitals use machines so that they can test a lot of blood samples quickly. Describe how you could check for anaemia using only a microscope.

....

....

....

1 Read the following paragraph and fill in the gaps with words from the box below. Each word may be used once, more than once, or not at all.

Blood leaves your heart in .... and returns in .... The right side of your heart pumps blood through your .... to collect .... The left side sends blood to every other part of your .... in every tissue bring blood close to all your ....

oxygen	veins	lungs	body	cells	capillaries	arteries

2 The diagram below shows the human circulatory system.

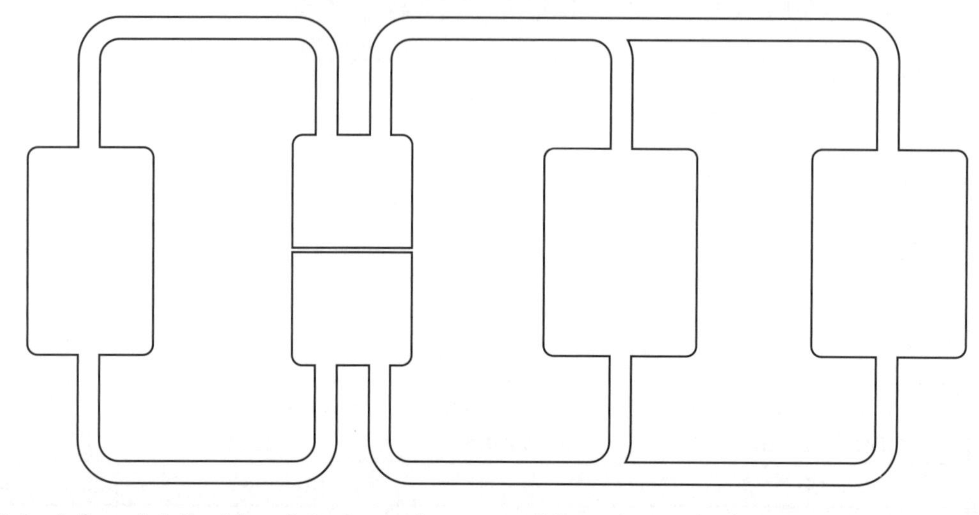

a Label the left and right sides of the heart, lungs, small intestine and other tissues.

b Colour all of the blood vessels that are high in oxygen in red, and those that are low in oxygen in blue.

c Label two veins entering your heart and two arteries leaving it.

3 The diagram shows how blood vessels divide to form capillaries in each of your organs.

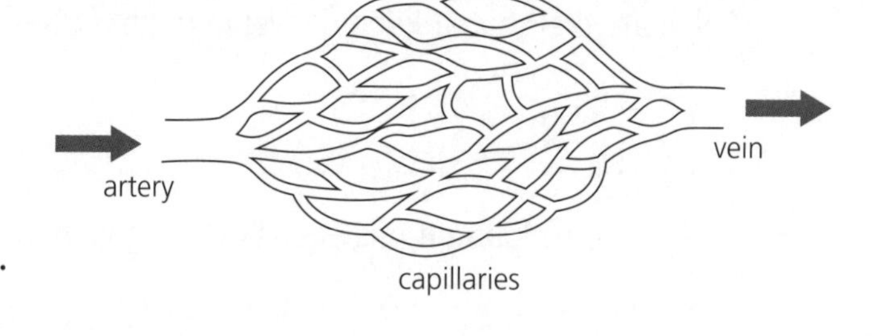

a Explain how blood changes as it passes through capillaries in your lungs.

....

....

b How does blood change as it passes through the capillaries around your small intestine?

....

....

E

The diagrams show open and closed valves. Describe one place where they are found and explain why they are needed.

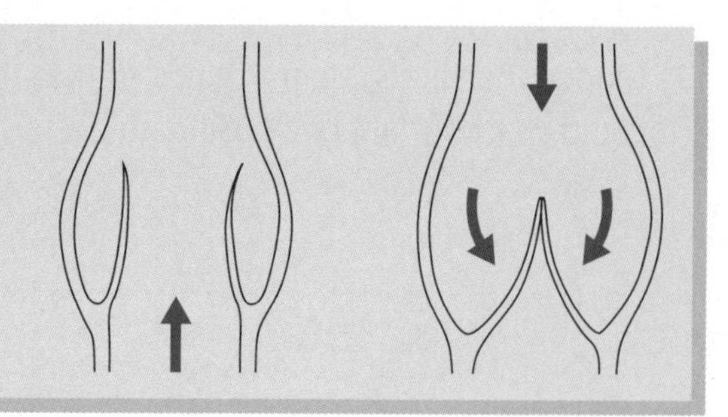

9.4 Identifying trends

1 Read the following paragraph and fill in the gaps with words from the box below. Each word may be used once, more than once, or not at all.

Your heart beats harder and .... during exercise. Regular exercise makes your heart more .... This .... your .... rate and .... your recovery ....

efficient	heart	time	faster	decreases	shortens

2 The table shows two athletes' test results.

a Deduce which athlete is fittest..................

b Give two pieces of evidence to support your conclusion.

....

....

Athlete	Resting heart rate (beats per minute)	Recovery time (seconds)
A	63	125
B	74	457

3 The table shows how an athlete's cardiac output changed as their heart rate increased.

Heart rate (beats per minute)	Cardiac output (dm^3 per minute)
40	5
80	13
120	21
160	28

a Use graph paper to display these results.

b Describe the relationship between the athlete's heart rate and cardiac output.

....

c Explain why an athlete's heart rate needs to increase when they run faster.

....

....

E

Karis collected these results. Plot a graph to display the results.

a Describe the correlation between her heart rate and her blood pressure?

b Predict what her blood pressure would be if her heart rate was 80 beats per minute.

Heart rate (beats per minute)	Blood pressure (units)
82	118
63	106
68	108
86	116
73	114
90	121

1 The diagram shows a partially blocked artery.

a Label the diagram to show what is causing the blockage.

b Describe how blockages like this affect blood pressure.

....

2 The following statements describe how a poor diet can lead to a heart attack. Write the letters in the correct order on the line below. Start with statement C.

....

A Heart muscles die.

B His blood pressure increases.

C The patient eats food high in saturated fat and salt.

D The plaque blocks an artery that supplies the heart with oxygen.

E Plaque begins to form in his artery walls.

F Pain spreads across the patient's chest and left arm.

G Some of the plaque breaks away from an artery wall.

H The patient may die.

3 Three patients were asked to record what they ate.

Meal	Patients		
	A	B	C
Dinner	fish, vegetables	processed food	chicken, rice
Snack	fruit	biscuits	nuts
Drink	fruit juice	cola	cola

Which patient's diet increases their risk of having a stroke? ....

E

Scientists compared the importance of different heart attack risk factors. The bar chart shows some of their results.

Write down three things an office worker could do to reduce their risk of heart disease.
Put the most important thing first.

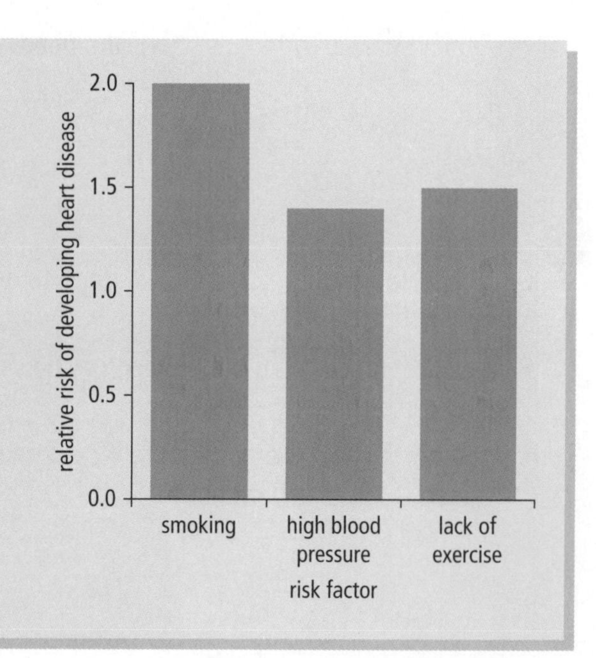

10.1 Lungs

1 Add labels to the diagram below to show how we breathe in and out.

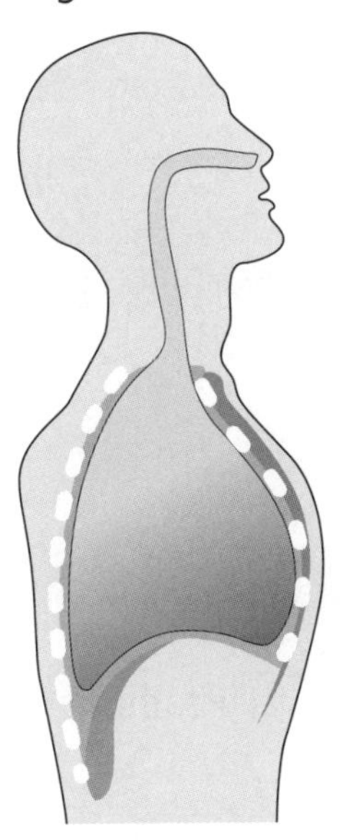

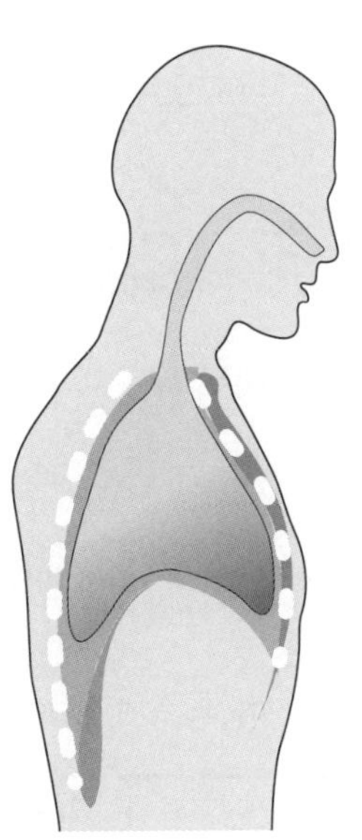

2 Label the diagram below. Then add arrows to show how air travels to the alveoli.

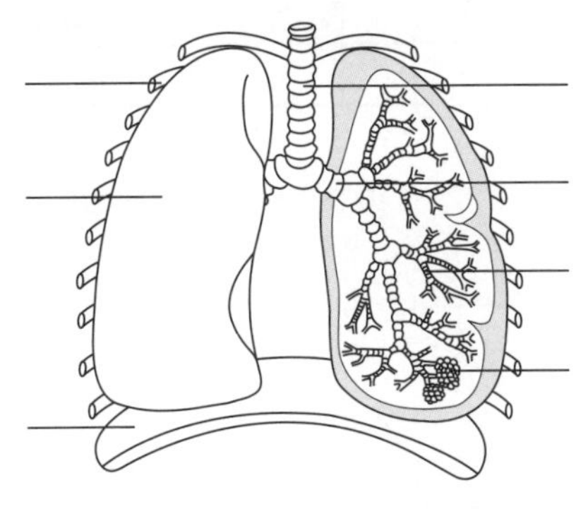

3 Read the following paragraph and fill in the gaps with words from the box below.

Your diaphragm, and .... between your ...., make air move in and out of your .... It travels through the trachea, bronchi, and .... to .... In the alveoli, oxygen .... into the blood and carbon dioxide .... This is gas ....

alveoli	ribs	diffuses	leaves	bronchioles	exchange	lungs	muscles

E Explain why alveoli allow rapid diffusion between blood and air.

10.2 Respiration and gas exchange

1 Fill in the gaps in the following paragraph with words from the box below.

Gas exchange means taking oxygen into your .... and releasing carbon dioxide. It happens in your .... Respiration is the release of .... from food molecules such as .... It happens in all your .... Respiration speeds up during .... to provide your muscle cells with extra energy. Gas .... also speeds up to provide the extra .... needed for respiration.

exercise oxygen glucose exchange blood cells alveoli energy

2 Complete the table to show the difference between breathed in and breathed out air.

Gas	Breathed in air (%)	Breathed out air (%)
Oxygen	21	
Carbon dioxide	0.03	
Nitrogen and other gases	79	79

3 The spirogram shows how the volume of air in a student's lungs changes as she breathes in and out.

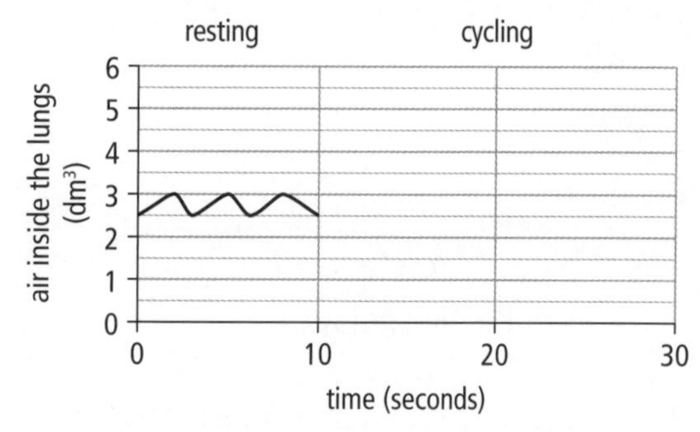

a Calculate her breaths per minute.

....

b What volume of air does she take in with each breath? ....

c Continue the graph to show how her breathing changes as she starts to cycle.

E

The *ventilation rate* is the volume of air that enters the lungs per minute. It is calculated by multiplying the volume of each breath by the breaths per minute. The table shows how training at a high altitude affected one athlete's ventilation rate.

Describe what happens during training at a high altitude to cause this effect.

Activity	Average ventilation rate (dm^3/minute)	
	Before training at a high altitude	After training at a high altitude
Standing	12.5	10.2
Running	128.3	114.1

10.3 Anaerobic respiration

1 Complete the equations for aerobic and anaerobic respiration.

Aerobic: glucose + ____________ → ____________ + ____________

Anaerobic: glucose → ____________ ____________

2 The diagram shows an athlete's oxygen uptake before, during and after exercise.

a Add labels to show what is happening in each part of the graph.

b Explain why anaerobic respiration cannot be used all the time.

....

....

oxygen uptake

time

c Explain why anaerobic respiration can produce a sudden burst of energy even though it only releases a small percentage of the energy in glucose.

....

....

3 Write ***T*** after the true statements and ***F*** after those that are false. Then write corrected versions of the statements that are false.

a Anaerobic respiration does not require oxygen.

b Aerobic respiration only provides short bursts of energy.

c Anaerobic respiration produces lactic acid.

d Anaerobic respiration releases the same percentage of the total energy in glucose as aerobic respiration.

e Anaerobic respiration is the main type used in marathons.

Corrected versions of false statements:

....

....

....

4 Statements a–g refer to muscle fibres. Write the letter of each statement in the correct part of the Venn diagram.

a Specialise in aerobic respiration.

b Specialise in anaerobic respiration.

c Produce large forces.

d Contain more mitochondria.

e Take in oxygen at a faster rate.

f Can contract for longer periods of time.

g Can only contract for a short period of time.

h Contract to pull on bones.

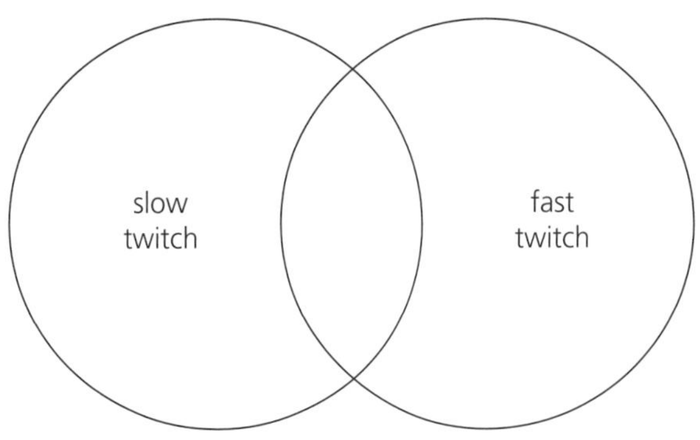

10.4 Smoking and lung damage

1 The diagram below shows the cells that line your lungs.

a Add labels to show how lungs are normally kept clean.

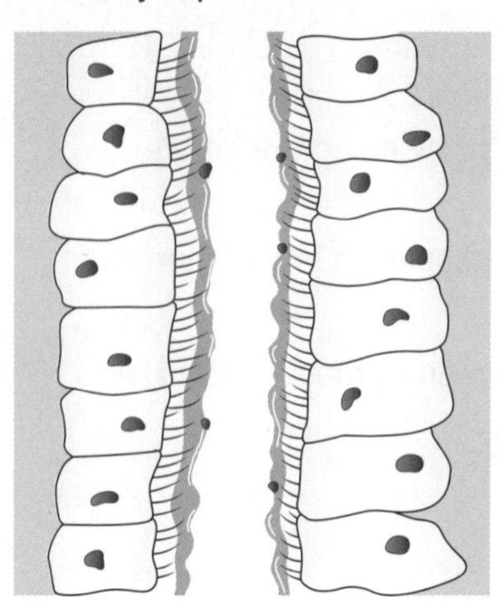

b List six changes that take place if you smoke.

....

....

....

c List two diseases that smokers are more likely to suffer from.

....

2 Fill in the gaps in the following paragraph with words from the box below.

Cigarette smoke contains many .... chemicals. Tar paralyses .... and makes mucus build up in the lungs. It is also carcinogenic so it can cause lung .... Carbon monoxide .... the amount of oxygen blood can carry. Nicotine .... smokers' blood pressure by causing blood vessels to become ...., so it increases their risk of having a .... attack. It also makes cigarettes ....

raises	cancer	addictive	narrower	harmful	reduces	heart	cilia

E

Two friends run a 200 m race. Neither do much exercise and one smokes 20 cigarettes per day. Their results are shown in the table.

Runner	Race time (seconds)
Imran	29
Mohamed	45

Identify the smoker and explain why his race time is so different.

10.5 Communicating findings

1 Fill the gaps in the following paragraph with words from the box below.

It is important to communicate findings .... Graphs, models, and ....

can make .... easier to understand. We can keep explanations simple by leaving out

.... that are not ....

details	clearly	relevant	explanations	pictures

2 Explain the following scientific terms.

a Peak Expiratory Flow (PEF)

....

....

....

b Asthma

....

....

....

....

3 Two students breathe out as fast as they can. A machine measures the volume of air they breathe out. The results are shown on the graph.

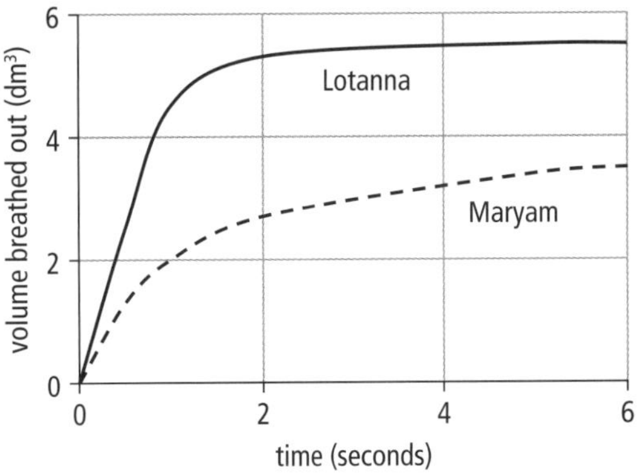

a Describe two differences between Lotanna and Maryam's results.

....

....

b Deduce which student has asthma.

....

volume breathed out (dm^3)
6
4
2
0
0
2
4
6
time (seconds)
Lotanna
Maryam

E

A doctor explains why smokers get more lung infections.

"The mucus in your bronchi normally traps the pathogens you inhale. Cilia then waft the trapped pathogens up your trachea. The tar from cigarette smoke paralyses the cilia and leaves the trapped pathogens in your lungs where they can cause infections."

a Write one reason why this explanation is unsuitable for patients with no scientific knowledge.

b Rewrite the explanation to make it clearer.

11.1 Reproduction

1 Fill the gaps in the following paragraph with words from the box below.

A new life begins when a sperm nucleus .... with an egg cell .... in an oviduct. This is .... The egg is released from an ovary and swept along the .... towards the uterus. Sperm are made in the ...., and pumped out through the sperm duct and .... They swim from the .... to the oviduct to find the egg.

fuses	testes	penis	fertilisation	vagina	nucleus	oviduct

2 Match each part of a male or female reproductive system to its function.

Body part
Ovary
Oviduct
Vagina
Uterus
Testes
Sperm duct
Penis

Function
Where fertilisation takes place
Where sperm enter a woman's body
Where egg cells develop
Sperm is produced here
Used to place semen in a woman's vagina
Where the embryo implants and develops
Carries sperm past glands that add fluids to form semen

3 Label the male and female sex cells to show how they are suited to the jobs they do.

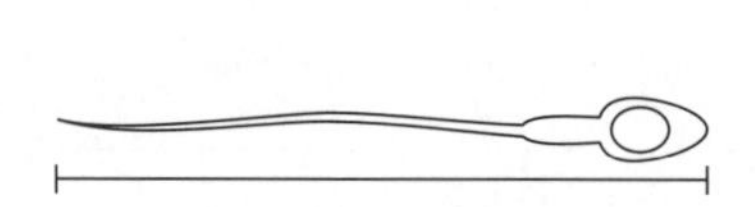

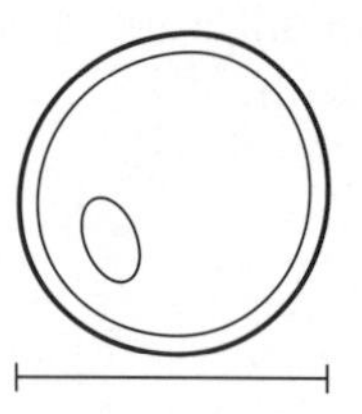

E Use diagrams to explain how two sex cells can join to form a new life.

11.2 Fetal development

1 Put the following sentences in order to show how a fetus develops. Start with statement G.

The correct order is:...

A The embryo becomes a hollow ball of cells.

B The fetus has a full set of sense organs.

C The fetus is mature enough to survive outside the uterus.

D The fertilised egg cell begins to divide to form an embryo.

E The embryo's cells form tissues and organs and it becomes a fetus.

F The embryo implants in the wall of the uterus and develops a placenta.

G A sperm cell nucleus fuses with an egg cell nucleus during fertilisation.

2 The diagram below shows a fetus that is ready to be born.

a Label the diagram.

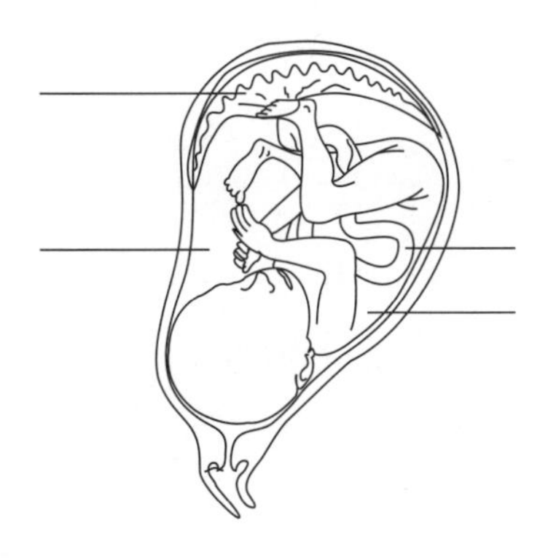

b Describe what the umbilical cord does.

....

c What use is the amniotic fluid?

....

d Explain why blood from the fetus needs to flow through the placenta.

....

....

3 The diagram represents the thin barrier between blood from the fetus and its mother's blood in the placenta.

Add labels to the arrows to show which way these molecules move: oxygen, carbon dioxide, glucose, urea.

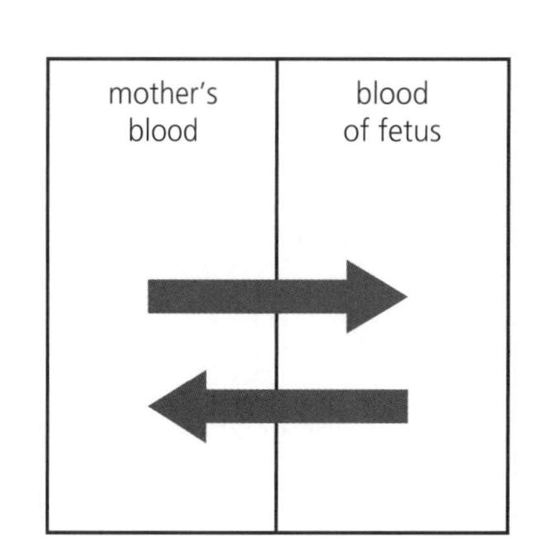

E Explain why pregnant women should not smoke or drink alcohol.

11.3 Twins

1 Fill the gaps in the following paragraph with words from the box below.

Non-identical twins form when two .... fertilise .... eggs.

Identical twins form when embryos .... into two. Conjoined twins form when

.... twins do not .... completely. Conjoined twins can be

.... if they don't share too many ....

split	separate	separated	sperm	organs	identical	two

2 Draw lines to match each observation to the question it prompted.

Observation
Twins are more likely to die in their first month than single babies.
On average twins have lower birth masses than single babies.
About one-third of babies die if their birth mass is below 1 kg.

Question
Could the birth mass of babies affect their survival?
Could more small babies be dying because they are born too soon?
What could be different about babies that are twins?

3 Babies that are born too early need to be kept warm in an incubator.

a The bar chart shows data about a hospital's use of incubators. Calculate the percentage difference between twins and single babies needing incubators.

....

b Suggest a reason for this difference.

....

....

c The babies kept in incubators often cannot suck or breathe independently. Describe how they are helped to survive.

....

....

....

11.4 Adolescence

1 Fill the gaps in the following paragraph with words from the box below.

During puberty, sex .... cause physical and .... changes in boys and girls. Their height increases .... and their .... organs develop. This is also the time when a girl's .... start. .... usually occurs between the ages of 11 and 15.

emotional	periods	puberty	rapidly	hormones	sex

2 Write ***B*** after changes that take place in boys, and ***G*** after those that take place in girls. Some take place in both boys and girls.

a Their voices get deeper.

b They undergo a period of rapid growth.

c Their hips get wider.

d Hair grows in their pubic region.

3 The diagram shows how the lining of a girl's uterus changes during her menstrual cycle.

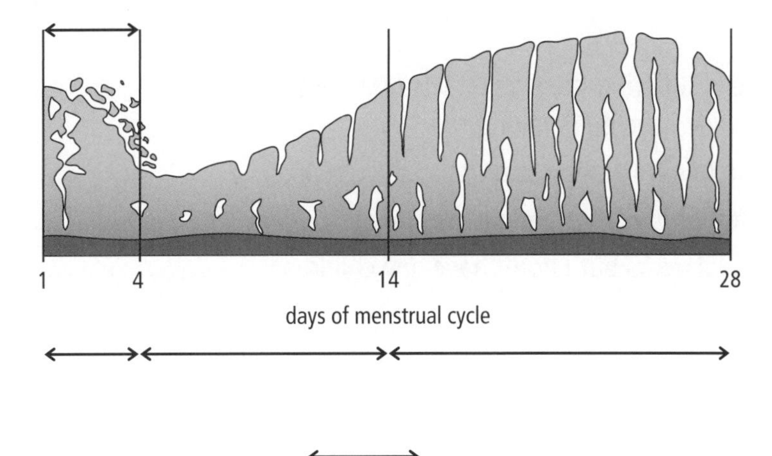

a Add labels to the diagram to show what is happening in each part of the cycle.

b Estimate the time when making love is most likely to lead to fertilisation.

....

c Write two reasons why a woman's periods may stop.

....

....

E

During puberty changes take place in many organs at the same time.
Explain how your brain makes all these changes happen at the same time.

1 There are two sorts of drugs: pharmaceutical drugs and social drugs. Decide which each statement describes and add its letter to the correct section of the Venn diagram.

a Used to treat diseases.

b Painkiller or antibiotics.

c Can be dangerous in high doses.

d Affect the way your body works.

e Always affect your nervous system.

f Caffeine, nicotine or alcohol.

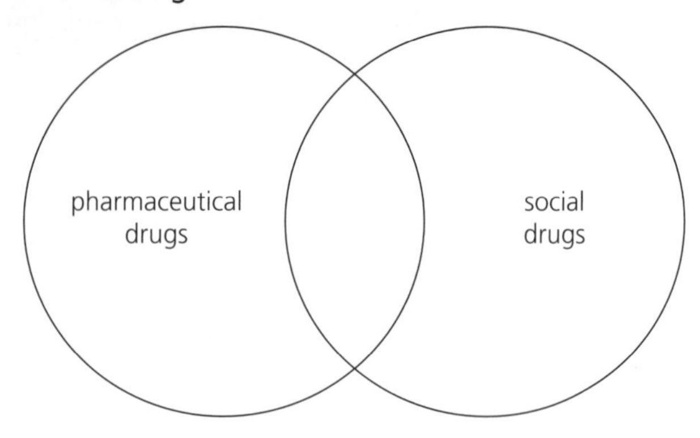

2 Draw lines to link each drug to the correct type and effects.

Drug
Caffeine or nicotine
Cannabis
Alcohol

Type
Hallucinogen
Depressant
Stimulant

Effects
Slows reactions. Increases accidents.
Speeds up reactions Makes users feel alert.
Changes the way users see and hear things.

3 Each of these people is suffering the side effects of a drug. Suggest which drug they have been taking.

a I find it impossible to get to sleep at night……………………………………………

b I keep getting into fights when I am out with friends…………………………………

c I've been married for years but I don't have any children…………………………....

E

Use the information about the following drugs to classify them as stimulants, depressants, addictive drugs or hallucinogens. Some drugs go in more than one category.

a Barbiturates are used in sleeping pills. These can be hard to give up.

b Cocaine makes people feel more energetic but the effect gets weaker each time users take it. If they keep taking more, they may die from an overdose.

c Ecstasy increases users' energy levels, makes them feel closer to other people and makes them more sensitive to touch.

1 Draw lines to link each disease to the organisms that cause it and the symptoms.

Disease
Bilharzia
River blindness
Tuberculosis
Chlamydia or gonorrhoea

Organism responsible
Worm
Bacteria

Symptoms
Blindness
Infertility
Anaemia, diarrhoea and liver damage
Fever and coughing

2 Doctors advise their patients to take these precautions. Name the disease each action will protect them from.

a Do not stay in the same room as someone with a persistent cough. ....

b Wear insect repellent and clothes that cover your arms and legs. ....

c Do not have unprotected sex. ....

d Make sure your food is not contaminated. ....

e Do not walk in infected lakes and ponds. ....

3 The World Health Organisation records the number of cases of tuberculosis every year. Use graph paper to display their results as a line graph.

Year	TB cases per 100 000 people
1990	270
1995	270
2000	265
2005	220
2010	175
2013	170

a How many cases were there for every 100 000 people in 1990? ....

b Calculate the fall in the number of cases between 1990 and 2013.

....

c Between which years was the fall in the number of cases most rapid?

....

E

Explain why chlamydia and gonorrhoea can only be passed to new victims during sex and during childbirth.

12.3 Defence against disease

1 Read the following paragraph and fill in the gaps with words from the box below.

Harmful micro-organisms are called .... Your .... keeps most of them out. Pathogens that invade your body are attacked by .... blood cells. Some white blood cells make .... These Y-shaped .... attach to pathogens and help white blood cells to .... them.

skin	destroy	white	proteins	antibodies	pathogens

2 Draw lines to link each component of your immune system with its role.

Defence
Skin
Stomach acid
Mucus
Antibodies
Phagocytes
Other white blood cells

Role
Traps the pathogens you breathe in
Proteins which help white blood cells destroy pathogens
Destroys pathogens in your food and drink
Keeps most pathogens out of your tissues
White blood cells which ingest pathogens
Make antibodies

3 The HIV virus destroys one type of white blood cell called a **helper T-cell**. The graph shows how the numbers of these cells changed in a patient with the disease.

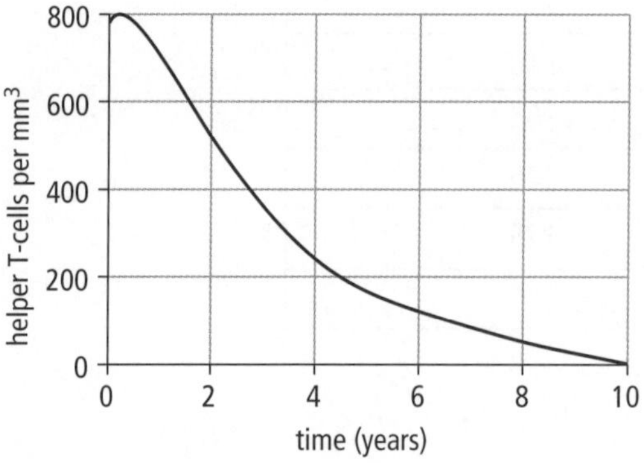

helper T-cells per mm^3: 0, 200, 400, 600, 800
time (years): 0, 2, 4, 6, 8, 10

a Describe how the number of helper T-cells changes.

....

....

....

b Ten years after becoming infected with HIV, the patient caught tuberculosis and died. Explain why they did not recover.

....

....

....

12.4 Boosting your immunity

1 Read the following paragraph and fill in the gaps with words from the box below.

Pathogens reproduce very .... White blood cells need to produce .... antibodies for each pathogen. This takes ...., so you may feel ill for a few days. If the same .... invades again, your white blood cells .... antibodies .... quickly so you don't get ill.

pathogen	more	different	quickly	produce	time

2 The graph shows how the number of antibodies in the blood changes during an infection. Add a second line to the graph to show what would be different if the same pathogen infected you a second time.

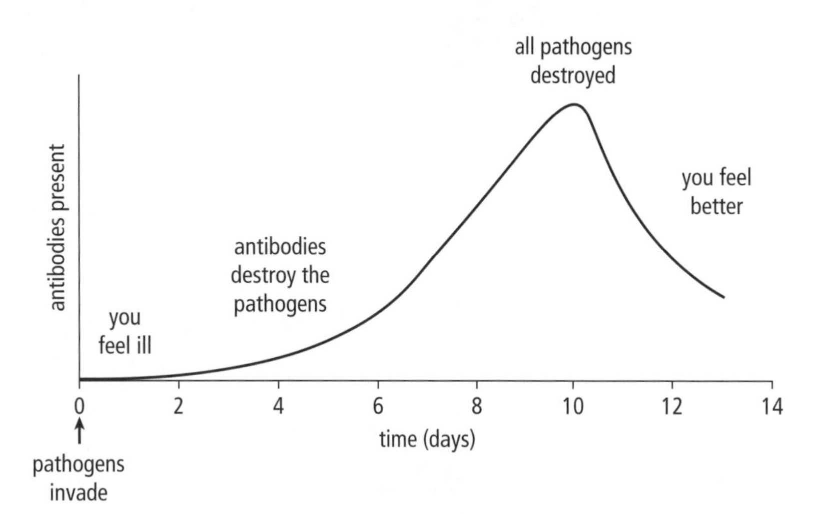

3 Decide whether each statement is true (***T***) or false (***F***). Write corrected versions of the false statements on the lines below.

a Vaccines contain dead or weak pathogens.

b Vaccines make white blood cells produce antibiotics.

c Vaccines stop you falling ill if the live pathogen enters your body.

d Once you have been vaccinated you are immune to any disease.

....

....

E

Put the following steps in order to explain how bacteria become resistant to antibiotics. Start with step C.

A Her doctor prescribes antibiotics.

B The toughest bacteria continue to multiply.

C A patient is seriously ill. She is infected with bacteria.

D The first few doses of antibiotics kill the weakest bacteria.

E Their resistance to the antibiotic passes to future generations.

F The patient feels better and stops taking the antibiotics.

G The antibiotic is no longer effective.

1 Complete the equation for photosynthesis.

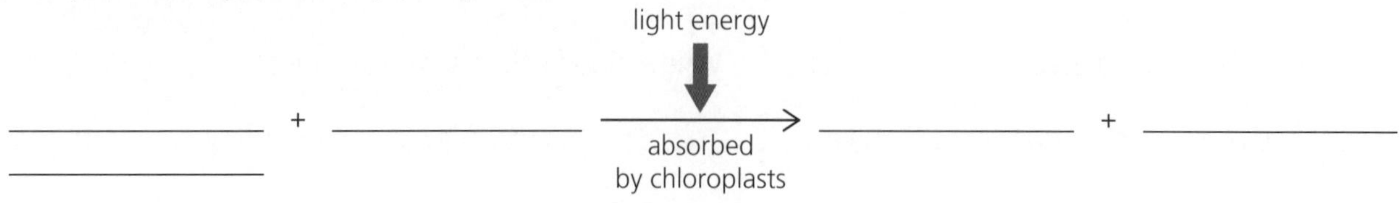

2 Add labels to the diagram below to show how energy absorbed by plants keeps animals alive.

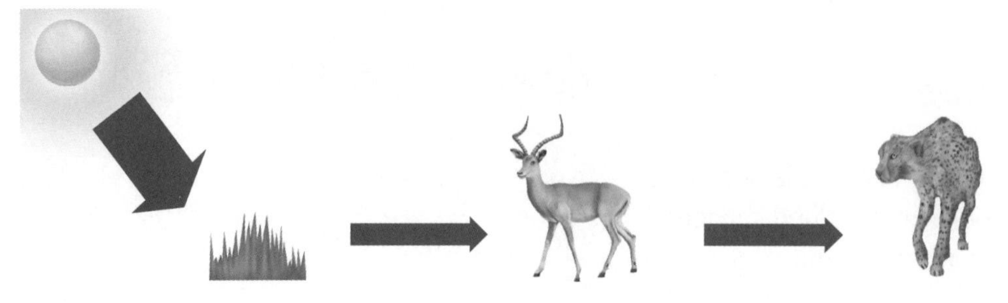

3 Add labels to the diagram below to show how water and carbon dioxide enter a leaf, where photosynthesis takes place, and what happens to the products made.

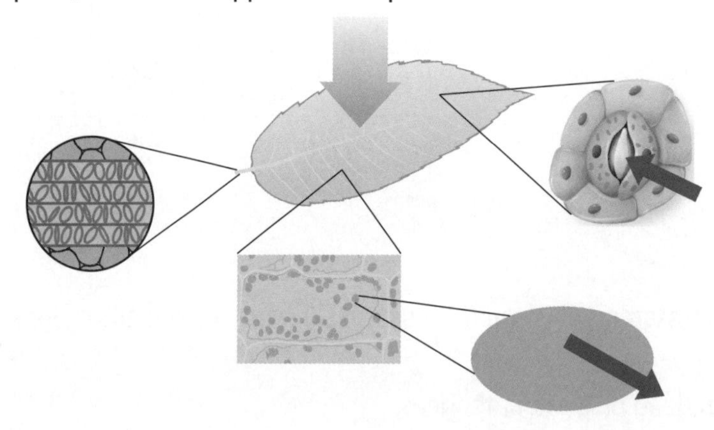

4 Abiola placed cress seeds on damp cotton wool in two dishes.He placed one dish in a dark cupboard and one in bright light.Draw the seedlings in the diagrams below.

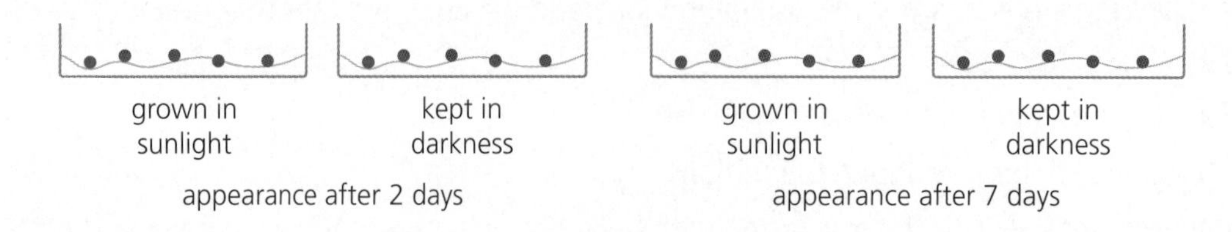

E

Aphids feed by pushing pointed mouthparts into the phloem tubes in a plant's veins.

a Explain what phloem tubes are and why plants need them.

b Suggest how the aphids affect the plants they feed on.

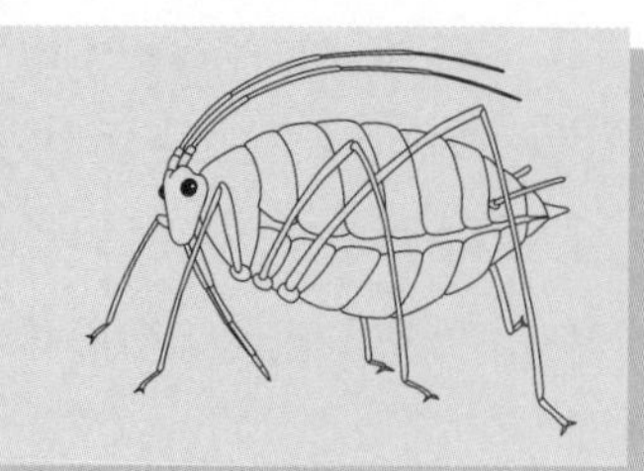

13.2 Preliminary tests

1 The four sentences below describe how Salma investigated photosynthesis. Write the letter at the start of each sentence in the correct box on the diagram.

a Salma measured the rate of photosynthesis under red, blue and green light.

b Salma wondered whether the colour of light affected photosynthesis.

c The results showed that photosynthesis was faster under red and blue light.

d Salma thought that plants could not use green light because they look green. This suggests that they reflect green light and absorb red and blue.

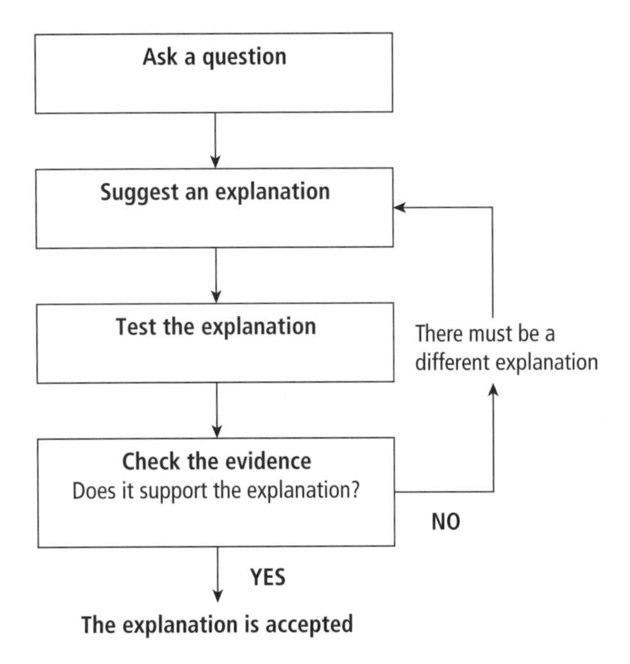

2 Link each piece of equipment Salma used to the job it does.

Equipment
Lamp
Filter
Syringe
Timer
Water

Job
Removes every colour from light except one.
To measure how long leaf discs take to rise.
Provides dissolved carbon dioxide for photosynthesis.
Produces white light.
Removes the air from leaf discs.

E

Marc plans to repeat Salma's experiment with a different plant. He checks his method.

a Leaf discs 50 cm from a lamp (in white light) take 47 minutes to rise. How should he change his method?

b Explain why this will improve his investigation.

c Marc tests his filter with a light meter. The reading under the green and blue ones is 100 lux but under the red one it is 200 lux. How should he change his method?

d Explain why this will improve his investigation.

1 Colour the leaves below to show how mineral deficiencies affect maize leaves.

phosphorous deficiency | potassium deficiency | nitrogen deficiency

2 Draw lines to link each statement on the left to one of the stages in developing a scientific explanation on the right.

What scientists did to learn more about a mineral deficiency
Compared the control and test batches of maize. The leaves only turn yellow in the test batch.
Suggested that the plants might be short of magnesium because the green chlorophyll in chloroplasts contains magnesium.
Wondered what makes maize leaves turn yellow in some parts of the country.
Grew two lots of maize. Gave the control batch every mineral and the test batch every mineral except magnesium.

Stage in developing a scientific explanation
Ask a question about something that has been observed.
Use creative thought to suggest a possible explanation.
Collect evidence to test the possible explanation.
Check the evidence to see if it supports the suggested explanation.

3 Complete the sentences below to describe why each mineral is needed.

Nitrogen is used to make ....

Magnesium is used to make ....

Photosynthesis and respiration rely on .... and ....

E

Scientists compared maize grown using hydroponics with maize plants grown using aeroponics. They measured how quickly they could take in minerals. The graph shows their results.

a Describe the results in the graph.

b Describe the main difference between hydroponics and aeroponics.

c Aeroponics lets roots take in more oxygen. Why is this gas needed?

d How would the differences shown in the graph affect the growth of the plants?

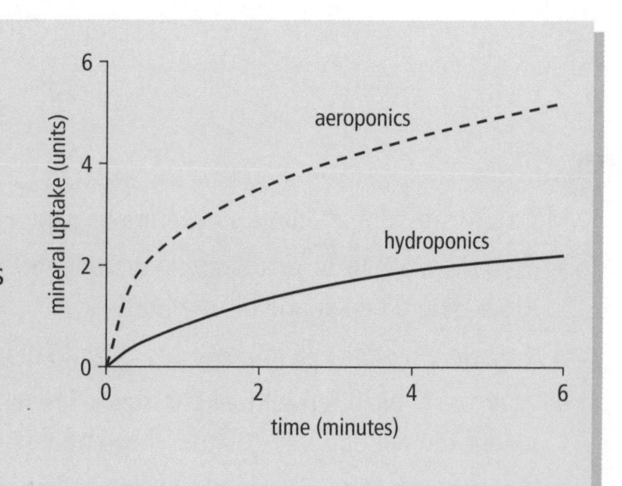

13.4 Phytoextraction

1 Make up six sentences about plants and metals. Each sentence must use one word or phrase from each column of the table. You may need to use some words or phrases more than once.

Plants need metals
Only small amounts
Many metals
High metal concentrations
Hyperaccumulators

prevent
can
to
are

absorb a lot of metal.
toxic in large quantities.
build their cells.
plant growth.
needed.
used to clean soil.

....

....

....

....

....

....

2 Colour and label the diagram below to explain how a hyperaccumulator can remove toxic metals from soil.

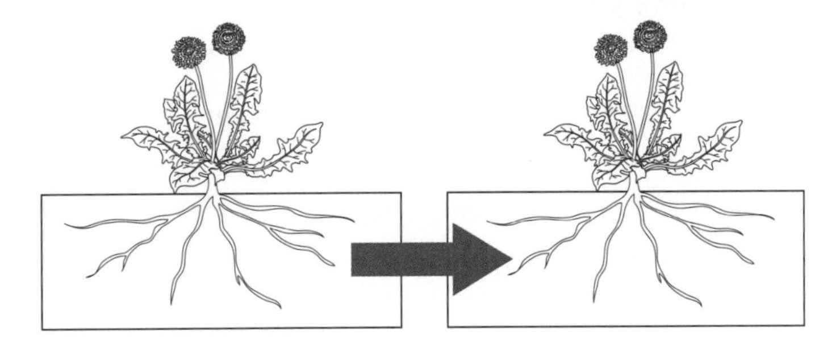

3 Scientists planted Chinese brake ferns in pots of contaminated soil. The graph shows how the amount of arsenic in these plants changed.

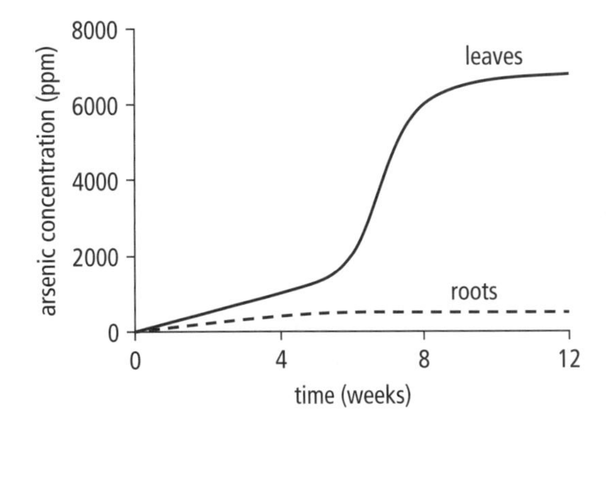

a Where was most of the arsenic accumulated?

....

b When did the ferns take in most arsenic?

....

c Scientists plan to burn the ferns. The arsenic will be left in a small quantity of ash. Suggest one thing they could do with the ash.

....

....

1 Label the diagram below to show the main parts of an insect-pollinated flower.

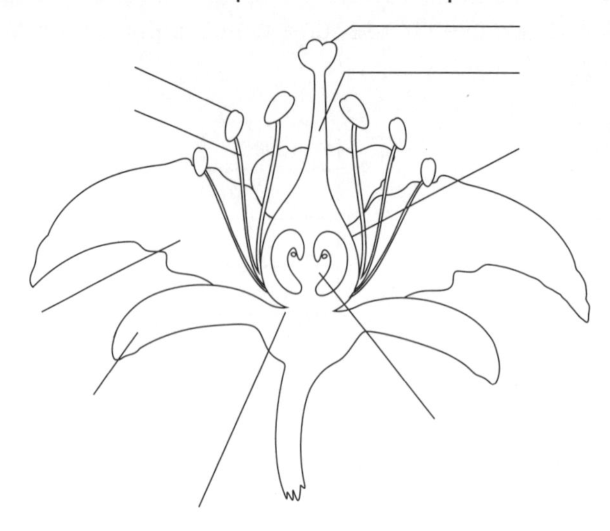

2 Name each of the following parts:

a Where ovules containing female sex cells are produced ....

b Where pollen containing male sex cells is produced ....

c Where the pollen has to land for pollination to take place ....

d The structure a pollen tube grows down to reach an ovule ....

e The part that forms a seed after fertilisation ....

f The part that forms a fruit after fertilisation ....

3 Label the diagram on the right to show the events that take place between pollination and fertilisation.

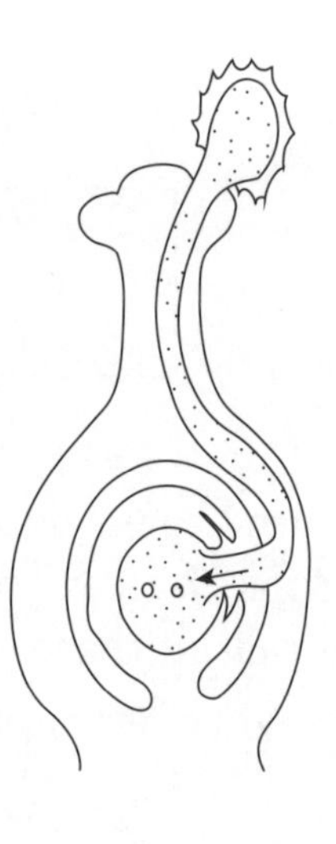

4 Label the diagram below to show the main parts of a wind-pollinated flower.

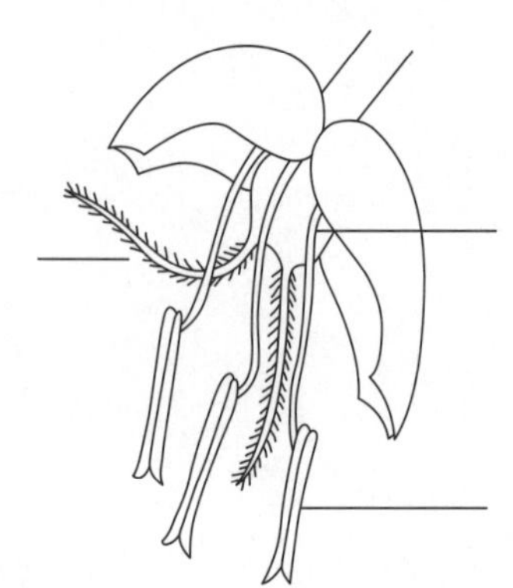

E

Many plants produce male and female sex cells in the same flower. They can produce seeds by self-pollination or cross-pollination.

a Describe the difference between these two processes.

b Explain why offspring show more variety when cross-pollination takes place.

13.6 Seed dispersal

1 Read the following paragraph and fill in the gaps with words from the text box below. Each word may be used once, more than once, or not at all.

Seeds can be spread by .... seed pods, wind, water or .... The

dormant .... inside a seed can survive extreme .... and lack

of .... They can travel long distances before they ....

embryo	exploding	animals	germinate	water	temperatures

2 Label each of the diagrams below to show whether the seeds are spread by exploding pods, wind, water or animals.

a ....

b ....

c ....

d ....

e ....

f ....

3 These diagrams show where seeds from two different plants were found.

a Which plant uses the wind to spread its seeds?

....

b Which plant's seeds are most likely to be spread by animals? ....

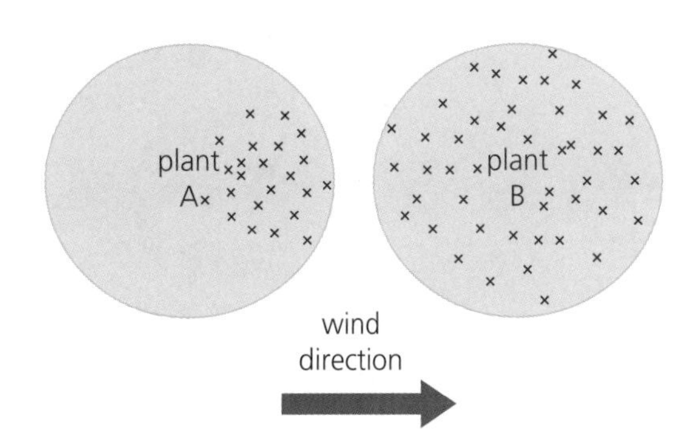

E

Cynthia compares three types of seed with wings. Each type of seed has a different mass. She holds each seed up to a fan and lets it go. She measures the distance the seed travels before it reaches the ground. She repeats the test 10 times for each seed type. The table shows Cynthia's results.

Average mass of seeds (g)	Distance travelled (m)		
	Less than 2	Between 2 and 4	More than 4
1	0	2	8
1.5	3	7	0
2	8	2	0

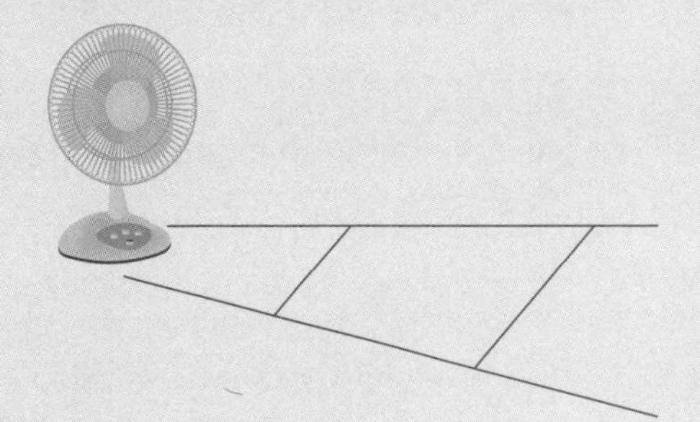

a What conclusion can you make from Cynthia's results?

b Explain why it is useful for seeds to get a long way from the parent plant.

14.1 Adaptation

1 Read the following paragraph and fill in the gaps with words from the text box below. Each word may be used once, more than once, or not at all.

Adaptations are .... that help an organism to .... in their usual .... Some adaptations are physical .... and others are .... Most adaptations take many .... to develop.

behaviours	generations	environment	features	characteristics	survive

2 Decide whether the following adaptations would be most useful to a **predator**, a **prey** animal or **both**.

a A hard shell ....

b The ability to run fast ....

c Sharp claws ....

d Camouflage ....

e Poisonous skin ....

f Eyes on the sides of their heads ....

g Tunnelling behaviour ....

h A smooth streamlined shape ....

3 These birds have adapted to eat different foods.

A B C

a Which bird crunches hard seeds? ....

b Which bird pulls worms out of mud? ....

4 Opossums hide in hollow trees during the day and feed at night. They eat almost anything – mainly smaller animals. They use their long whiskers to feel their way along the branches of trees. They have exceptional hearing and a good sense of smell.

a List two adaptations that help them avoid predators.

....

b List one adaptation that helps them find food.

....

E

Life on Earth has adapted to suit every possible habitat.

a Collect or draw images of five animals from your country.

b Label each animal to show the physical or behavioural adaptations that help it to find food or avoid predators.

14.2 Extreme adaptations

1 Draw lines to link each adaptation to desert life to the reason why it is useful.

Plant adaptation
Long roots near the surface
Swollen stems or leaves
Tiny leaves or spines
Long-lived seeds

How it saves water
prevents evaporation through stomata.
can survive for years until it rains.
store water for months between rain storms.
to catch occasional rain.

2 Read the following paragraph and fill in the gaps with words from the text box below. Each word may be used once, more than once, or not at all.

Large animals store more heat so they cool down more .... They can reduce heat loss by having a .... shape with a low .... A thick layer of fur traps air, which is a good .... Small animals can also use snow to trap a layer of .... around them. Layers of .... help to cut heat loss and act as an .... store.

air	surface area	fat	slowly	energy	rounded	insulator

3 Circle the animal from each pair which is best adapted to Arctic conditions.

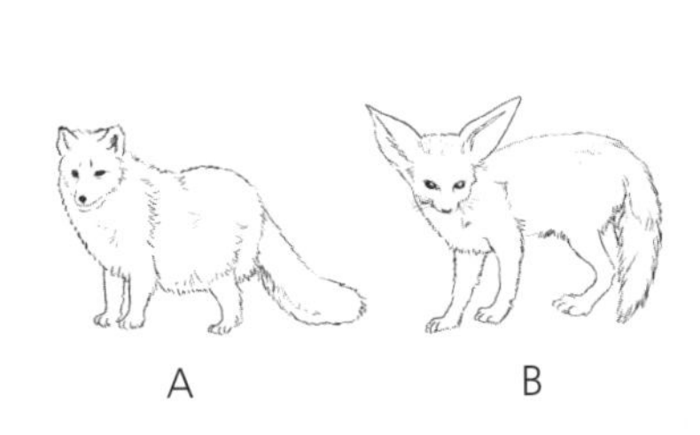

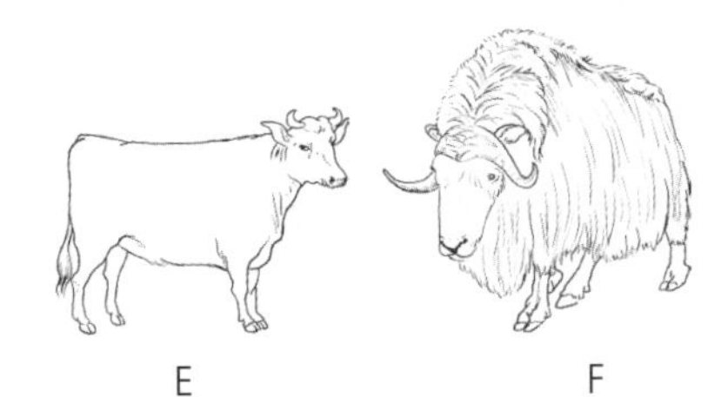

E

Deep under the oceans is one of the most extreme environments on Earth. The water is only a few degrees above its freezing point and there is no light.

In a few places superheated water gushes out of vents in the Earth's crust. The vents are called 'black smokers', because the water is full of minerals and these make it look black. Animals live near the vents that are found nowhere else on Earth. Two of these are the tube worm (*Riftia pachyptila*) and the pompeii worm (*Alvinella pompejana*).

Do some research. Then describe the adaptations that help each worm survive.

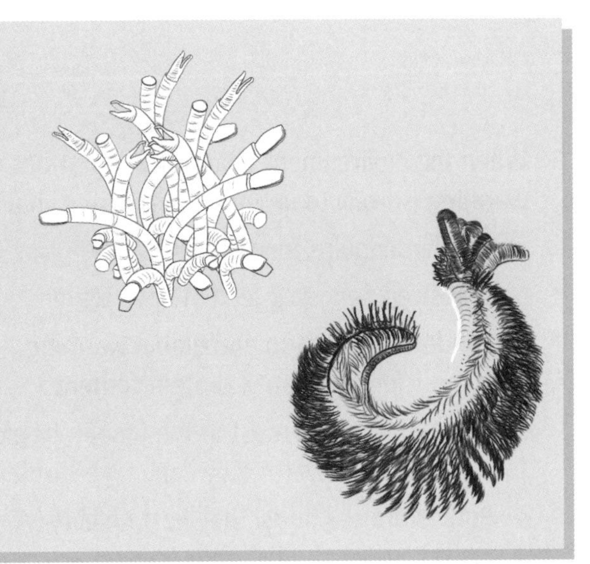

14.3 Survival

1 As global warming raises Earth's average temperature many species are migrating to places that better suit their adaptations.

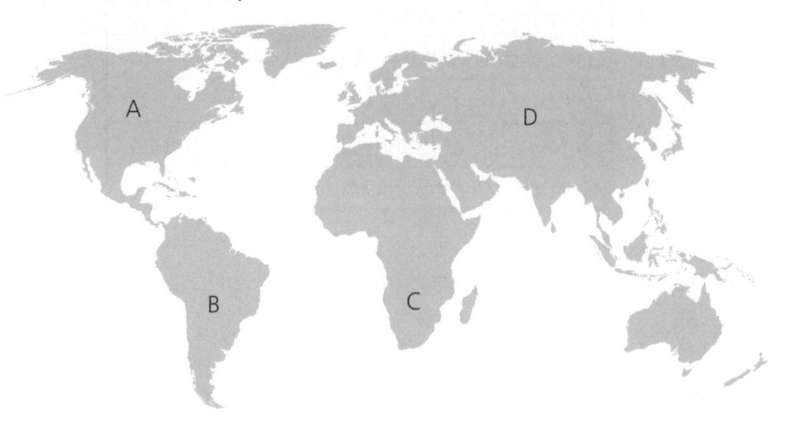

Add arrows to the map to show which way species from positions A–D are moving.

2 Earth's temperature is rising quickest at the poles. Every summer some of the sea ice in the Arctic melts. Scientists use satellites to measure the minimum area of ice left at the end of September.

a Display their results as a line graph on the axes below.

Year	Minimum area of sea ice (million m^2)
1978	8.2
1984	7.5
1990	6.9
1996	6.8
2002	6.8
2008	5.4
2012	3.6

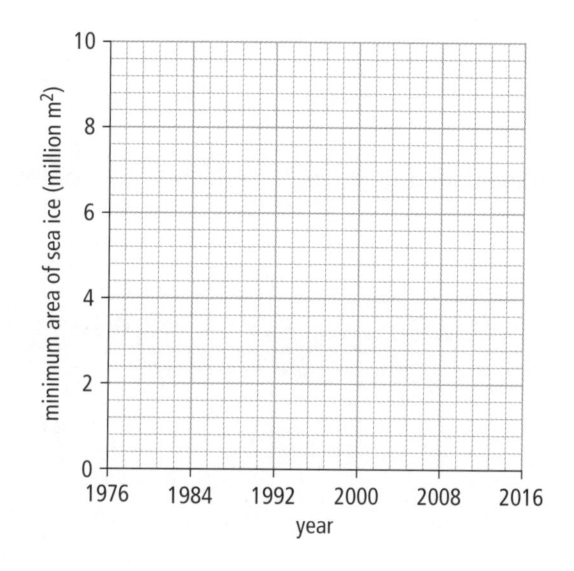

b Explain why the pattern on the graph makes it harder for polar bears to feed.

....

....

3 When the environment changes, adaptations can become more or less useful. Predict whether global warming is likely to **increase, decrease** or have **no effect** on the numbers of each animal.

a Adelie penguins feed under sea ice. ....

b Chinstrap penguins feed in open water. ....

c Pandas eat bamboo and global warming is reducing the plant's protein content. ....

d Sea turtles spend most of the time in tropical seas. As the oceans warm they can move further north. ....

e Puffins are sea birds. They nest on cliffs. As the oceans warm, the fish they feed on are moving north. ....

14.4 Sampling techniques

1 Read the following paragraph, and fill in the gaps with words from the text box below. Each word may be used once, more than once, or not at all.

As the human population has risen, the numbers of other .... have dropped.

Scientists .... how many plants or animals there are by .... their

populations. Plants are sampled using .... and animals are sampled using ....

Then an estimate of the total population can be ....

sampling quadrats species calculated traps estimate

2 Criminals often kill elephants so that they can sell their tusks. Since 1989, the sale of elephant tusks has been banned. Rangers regularly count the elephants to check that none are being killed. The table shows how their numbers have changed in the Serengeti National Park.

Year	Elephant numbers
1980	2200
1984	1800
1985	500
1989	550
1998	2000
2000	1600
2012	3000

Display these results as a line graph on the axes provided.

a Describe what the graph shows.

....

....

....

b Suggest reasons for these patterns.

....

....

E

Alake keeps spotting spiders in the field around her school. She decides to estimate how many there really are using the 'mark and recapture' method. The table shows her results. Use them to estimate the size of the population.

Number caught, marked and released on Day 1	15
Number caught on Day 8	12
Number of marked spiders recaptured	3

14.5 Studying the natural world

1 Read the following paragraph and fill in the gaps with words from the text box below. Each word may be used once, more than once, or not at all.

Scientists study animals to find out where they ...., what they .... and how they .... Some animals are hard to observe so scientists study their .... and the .... they leave behind. They also capture evidence using automatic .... and electronic ....

faeces	tags	live	cameras	behave	tracks	eat

2 Match each footprint to the animal that made it. Write the two correct letters next to each animal's name.

rhino................

impala..............

cheetah............

elephant...........

lion..................

A B C D E

F G H I J

3 Suggest one thing scientists can learn from automatic cameras that they cannot learn from studying animal tracks.

....

....

....

....

4 Suggest one thing scientists can learn from electronic tags that they cannot learn from studying animal tracks or using cameras.

....

....

E

Scientists often observe animals to find out more about the way they behave.
Choose an animal to observe. It can be a wild animal, a pet or a tiny invertebrate.
Describe what it spends most time doing, what it notices, and how it responds

15.1 Food webs

1 The diagram shows part of a desert food web.

a Add a producer to the food web.

b Name all the consumers.

c Underline all the primary consumers.

d Double underline all the secondary consumers.

e What type of consumer is a fennec fox?

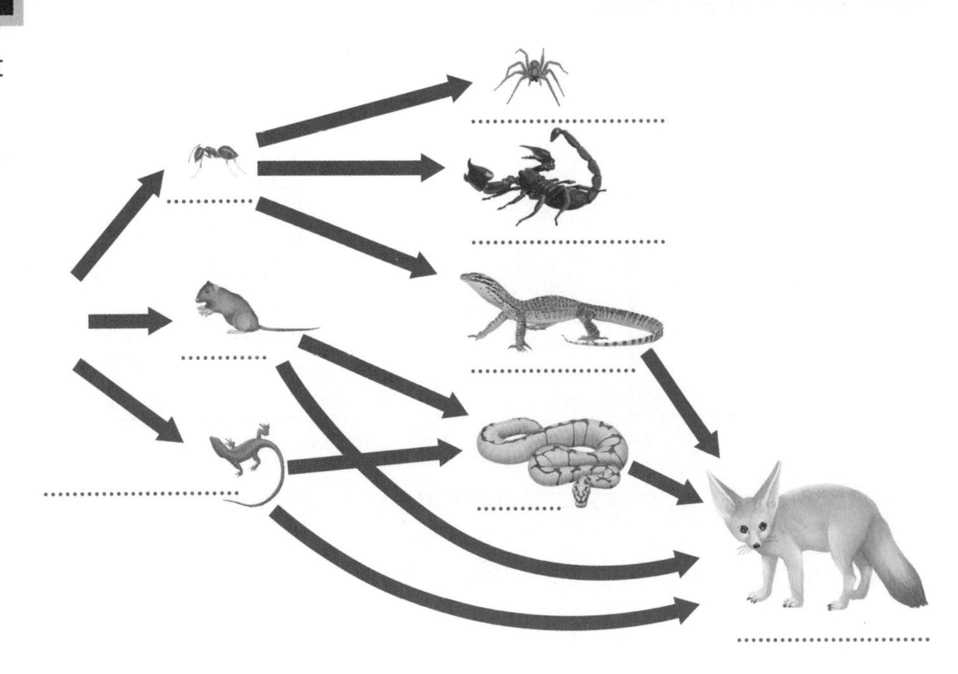

....

....

f Some lizard species are herbivores. Others are carnivores. Draw two food chains, one for each type of lizard.

....

....

g Complete the diagram on the right by putting each organism in each lizard's food chain in the correct trophic level.

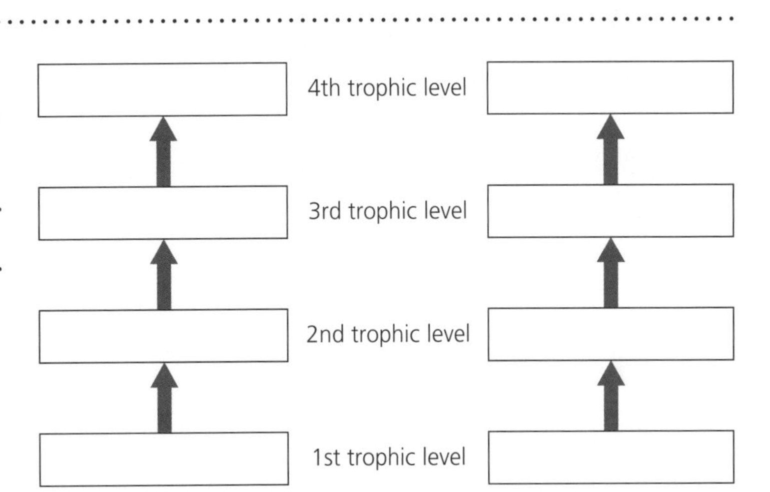

E

Marc made some observations in a habitat in North America. These are his notes.

Deer eat trees and shrubs, and rabbits eat shrubs and grasses. Mice and crickets just eat grass. These herbivores have many predators. Frogs eat crickets, mountain lions eat deer and rabbits, owls eat frogs and mice, and hawks eat rabbits and mice.

a Draw a food web for the habitat.

b Which animals feed at the second trophic level?

c Explain why it is difficult to assign a trophic level to owls.

15.2 Energy flow

1 Draw lines to link each word to the correct description.

Word
Food chain
Pyramid of numbers
Pyramid of biomass
Energy losses
Energy flow

Description
Shows the mass of living things in each trophic level.
The transfer of energy from one organism to another.
Shows where each organism gets its energy from.
Shows the number of organisms in each trophic level.
The energy organisms release during respiration or lose in the waste products they excrete.

2 Complete these sentences using the words in the box below.

Pyramids of number show the number of .... in each .... level but pyramids of .... show their total mass. This gets less at each trophic level because every living thing loses most of the .... they take in. Some is lost as heat during .... and some is lost in the waste .... living things excrete. Their consumers can only get the energy left in their ....

trophic	respiration	products	biomass	tissues	energy	organisms

E

Eniola investigated the energy flow from grass to a cheetah.

a The grass took in 10 000 J of solar energy and lost 5000 J as waste heat.

There were 3000 J of energy in its waste products. Add these values to the diagram and calculate the maximum amount of energy passed to impala.

b Impala lose 70% of the energy in their food as waste heat and 20% in their waste products. Calculate how many joules this is and add it to the diagram.

c Calculate the number of joules passed to the cheetah and add it to the diagram.

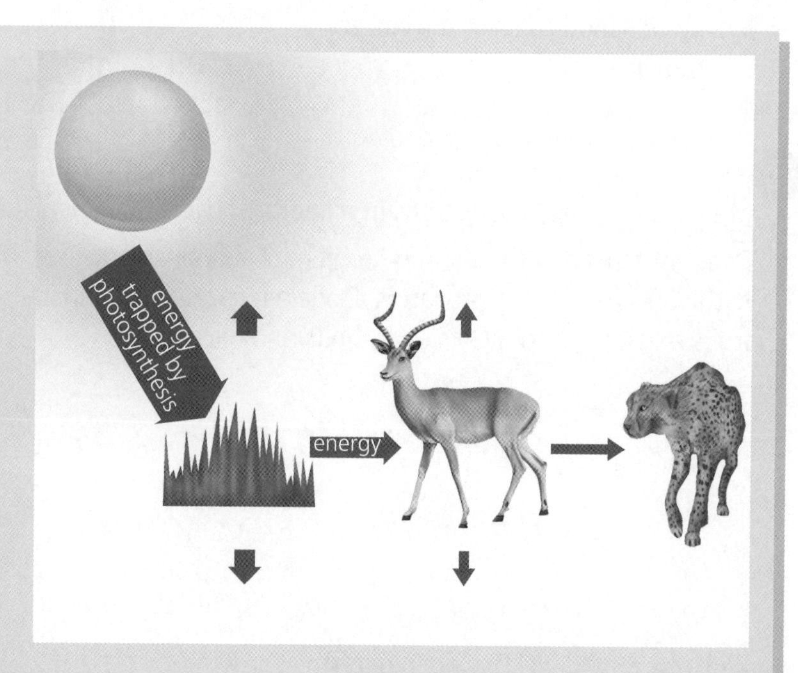

15.3 Decomposers

1 Fill the gaps in the following paragraph with words from the box below.

On land most plants are eaten by .... In mangrove forests, most plant material feeds decomposers. Decomposers break down organic waste to release .... Organic waste is made by .... things. It includes dead plants and .... and waste products like .... Decomposition is important because it returns .... to the soil. Most decomposers are fungi or ....

faeces	minerals	living	herbivores	bacteria	animals	energy

2 Put the living things in the diagram in the correct order to form a food chain.

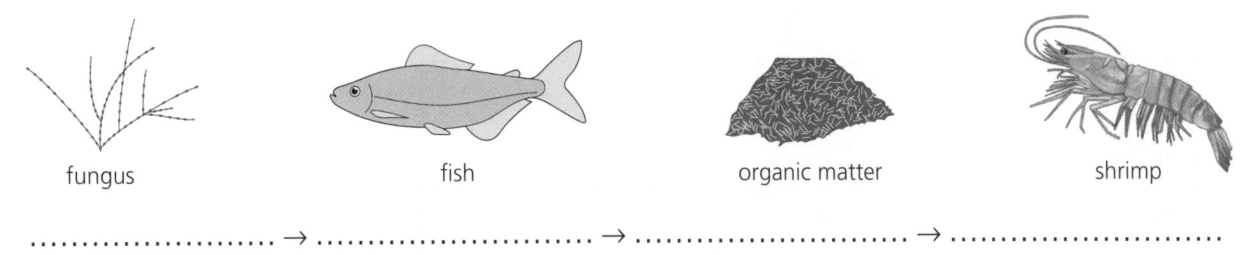

.... → .... → .... → ....

3 Add labels to the diagram below to show how carbon is recycled in a mangrove forest.

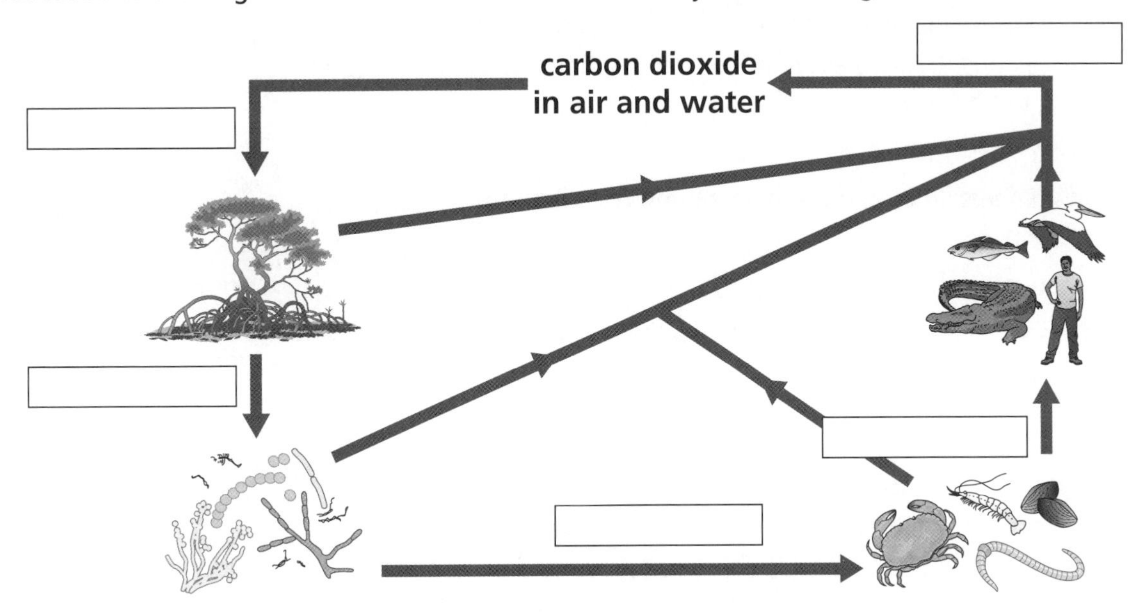

E

Scientists compared how much new biomass plants produced in neighbouring ecosystems in America. The results are shown in the table.

a Display the highest estimates as a bar chart.

b Give one reason why the mangrove forests are so productive.

Ecosystem	Average biomass produced per year (tonnes/hectare)
Desert	0.3
Farmland	1.5–5
Mangrove forests	5–15
Coastal water	1–1.5
Mid ocean water	0.3

15.4 Changing populations

1 Label the graph to show how populations usually change when animals move into new areas.

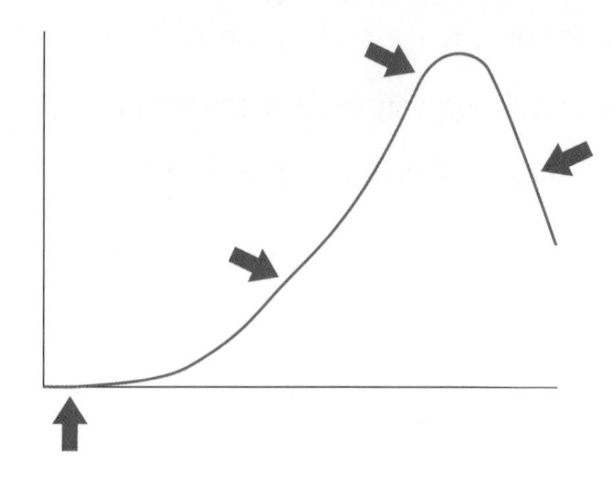

2 Match each word to the correct definition.

Word
Population
Interdependent
Sustainable
Biodiversity

Definition
Species which affect each other's numbers.
The number of individuals present.
A measure of the number of species present.
Able to continue for ever.

3 In parts of the Arctic the populations of caribou and wolves are interdependent. Add these labels to the graph below to show how this happens.

a Caribou eaten by wolves.

b Wolf numbers drop because there are fewer caribou to feed them.

c Caribou raise offspring.

d Wolves kill more caribou and breed, so their numbers rise.

e The cycle repeats.

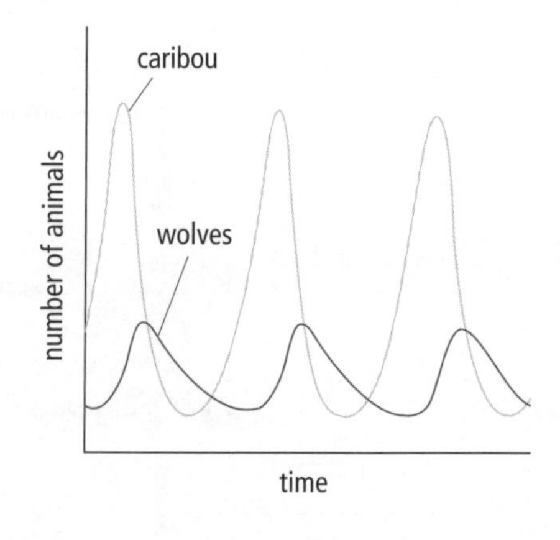

E

Large populations can damage the environment. In 1978, the Chinese government banned families from having more than one child. They claim that this policy has prevented more than 600 million births. The table shows how the Chinese population grew in the past and is expected to grow in the future.

Display the results on a graph and describe the pattern the graph shows.

Year	Population (millions)
1950	560
1960	650
1970	820
1980	980
1990	1150
2000	1270
2010	1350
2020	1430
2030	1460
2040	1450
2050	1420

15.5 Facing extinction

1 Fill the gaps in the following paragraph with words from the box below.

Population sizes are normally controlled by the amount of .... available, the number of predators and .... If new .... appear, or the food supply drops, a species can become .... Many extinctions are caused by .... species. These grow and reproduce .... and have few predators.

predators	quickly	extinct	food	invasive	disease

2 Brown tree snakes have caused drastic changes on Guam.

a Link these species to show the food webs present before and after their arrival.

Before:

lizards **insect-eating birds**

spiders

seed- and fruit-eating birds insects bats

leaves, fruit, and seeds from a wide range of plants

before

After:

brown tree snakes

lizards

spiders

insects bats

leaves, fruit, and seeds from a wide range of plants

after

b Write 'up' next to the species whose numbers went up, and 'down' next to the species whose numbers went down.

insect-eating birds..........
lizards..........
spiders..........
bats..........
seed and fruit-eating birds..........
plants..........

c How will the forest be affected if the insect population increases?

....

....

E

Insect-eating birds cut the number of insects feeding on trees. Seed- and fruit-eating birds spread their seeds and make them more likely to germinate. Both types of bird have become extinct on Guam. Predict, with reasons, which loss will have the biggest effect on the forest.

15.6 Maintaining biodiversity

1 Fill the gaps in the following paragraph with words from the box below.

Many different .... species grow in tropical rainforests. These produce a vast quantity of new .... each year and provide a wide variety of different .... When they are cleared to grow crops, ... plummets because most of these habitats are .... Species found nowhere else can become ....

plant	extinct	biodiversity	destroyed	habitats	biomass

2 Use arrows to make five sentences using a word or phrase from each column.

Most biodiversity is found
Farms usually
Infectious diseases
We store seeds
To prevent inbreeding in small populations scientists

to
have
in

a low biodiversity.
prevent useful plant species from becoming extinct.
produced offspring using sperm from a different part of the world.
warm, wet ecosystems.
destroyed food crops in the past.

3 Scientists have estimated how many vertebrates are threatened with extinction. The table shows how the number of threatened species has changed recently.

Vertebrate group	Threatened species (% of total in 2000)	Threatened species (% of total in 2011)
fish	3	7
amphibians	3	28
reptiles	4	8
birds	12	12
mammals	24	21

a On graph paper, plot a bar chart to display the data.

b Describe three differences between the vertebrate groups.

....

....

....

....

....

....

16.1 Air pollution

1 Label the diagram below to show where acid rain comes from.

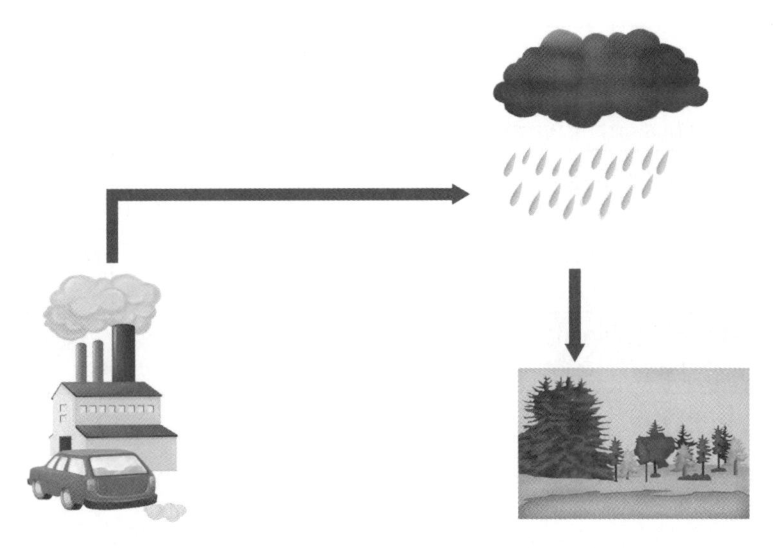

2 The diagrams show the main lichen species found in three different areas. Under each species, write what it tells you about the air at each location.

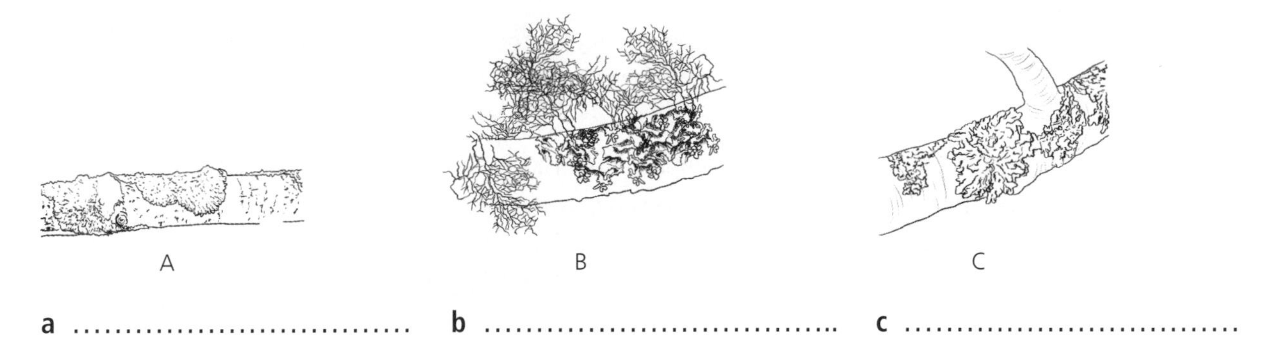

a .... b .... c ....

3 Write whether each process is 'increased' or 'decreased' by acid rain.

a Uptake of minerals from the soil ....

b The return of minerals to the soil ....

c The number of plants infected by pathogens ....

d The concentration of aluminium ions in lakes ....

E

American scientists collected primary evidence to show how the pH of lakes affects fish.

a Plot a line graph to show how pH affects the number of fish species present.

b Describe any patterns in the graph.

c Explain why it was important to have a lot of lakes in each group.

d Describe how flue gas desulfurisation and catalytic converters could prevent lakes becoming acidic in future.

pH value of water	Number of lakes tested	Average number of fish species present
4	7	1.3
4.5	92	1.9
5	118	3.0
5.5	110	3.6
6	154	4.8
6.5	240	5.8
7	297	6.0
8	96	5.8

16.2 How scientists work

1 Link each of the scientist's actions to a stage in developing a scientific explanation.

Scientist's actions
Joseph Fourier wonders why Earth is warm.
Joseph Fourier writes: 'the atmosphere must trap the Sun's heat.'
John Tyndall measures how much heat gases absorb.
John Tyndall finds that a small amount of carbon dioxide traps a lot of heat.

Stage in developing a scientific explanation
Suggest an explanation.
Test the explanation.
Ask a question.
Check the evidence.

2 Scientists use accepted explanations to make predictions. What prediction did Svante Arrhenius make?

....

....

....

....

3 Scientists used evidence from the past to compare the amount of carbon dioxide in the atmosphere with Earth's temperature. Complete the table to show what this evidence shows.

Measurements taken	What the evidence shows
Past temperature readings from weather stations all over the world.	
The thickness of tree rings, which show how fast trees grew each summer.	
The amount of carbon dioxide trapped in layers of ice, which shows how much was in the atmosphere.	

E

Scientists analyse ice cores to get carbon dioxide data from thousands of years ago.

a Use this data to make a line graph.(Make the x-axis run backwards from 100 to 0 thousand years ago.)

b Compare carbon dioxide levels now (time 0) and 20 000 years ago. How much extra is there now?

c Suggest why many scientists worry about the pattern shown in the graph.

Thousands of years ago	Carbon dioxide (ppm)
100	220
80	250
60	200
40	190
20	180
0	380

1 Complete the flowchart to show how eutrophication kills fish.

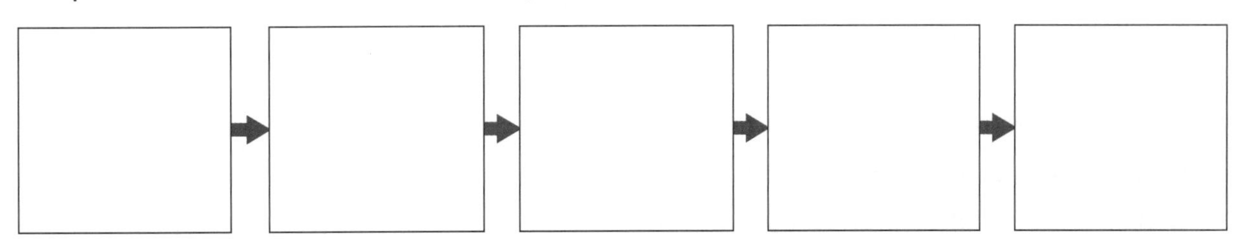

2 A large quantity of fertiliser has been washed into a stream. Decide whether the following 'rise' or 'fall'.

a Concentration of minerals in the water ....

b Total biomass of algae ....

c Light levels below the surface ....

d Amount of photosynthesis below the surface ....

e Mass of decomposing organic matter ....

f Number of bacteria ....

g Amount of respiration below the surface ....

h Amount of oxygen dissolved in the water ....

i Number of mayfly nymphs in the water ....

j Number of fish in the water ....

E

Scientists investigated bioaccumulation in a fish-eating bird's food chain.

Table 1 shows their results. Table 2 shows the results of a second investigation.

Table 1

Animal	Concentration of pollutant (ppm)
Fish-eating bird	25
Large fish	2
Small fish	0.5
Invertebrates	0.1
Water	0.000 01

Table 2

Concentration of pollutant (ppm)	Eggshell thickness (mm)
10	0.24
7	0.25
6	0.26
5	0.26
4	0.27

a Explain why the birds of prey contain so much of the pollutant.

b Display the results from Table 2 as a line graph and describe what it shows.

c Suggest why fewer birds are raising offspring successfully.

16.4 Saving rainforests

1 Decide whether each statement is a 'positive' or 'negative' effect of deforestation.

a It provides land to grow food or keep farm animals. ....
b It makes mudslides more common. ....
c It allows new roads to be built to connect cities. ....
d It can make the area warmer and windier. ....
e It can cause extinctions by destroying habitats. ....
f It allows more biofuels to be grown to provide energy. ....
g It increases the amount of soil carried away by rivers. ....
h It allows countries to earn extra money by exporting more products. ....
i It increases the amount of carbon dioxide in the atmosphere. ....
j It reduces the number of tourists who visit the area. ....
k It could increase global warming. ....

2 Add labels to the diagram below to show how reuse and recycling can cut pollution.

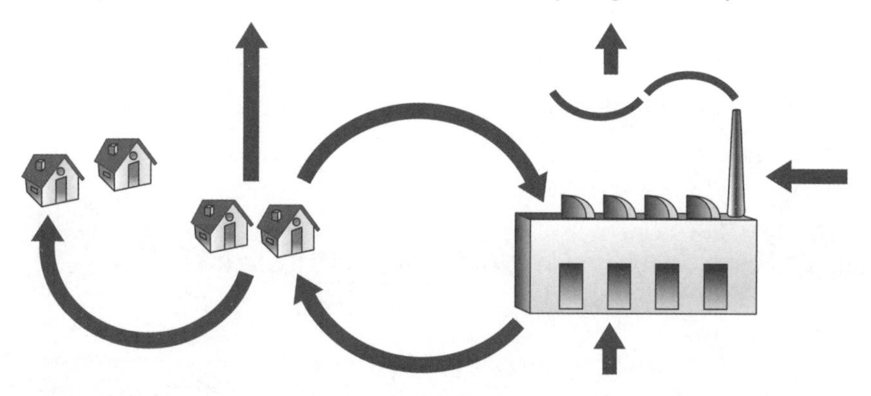

3 The bar chart shows how much energy and materials can be saved by recycling.

a For which resource is recycling most beneficial?.........................
b For which resource is the saving in materials much larger than the energy saving?.........................

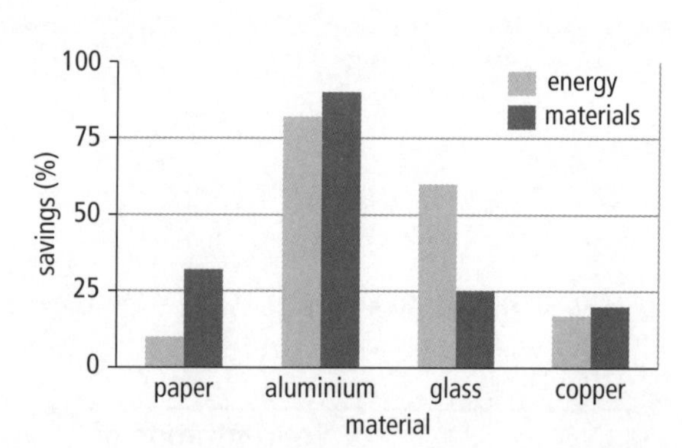

E

The table shows the continents with most of the world's tropical forests.

Continent	Forest cover (millions of hectares)	Percentage change since 2000
Africa	674	-10
Asia	593	3
South America	864	-9

a Which continent has the largest area of forest?
b Which two continents have lost most forest in recent years?

17.1 Using keys

1 Each of these snakes has hollow fangs. The fangs drop down when they open their mouths.

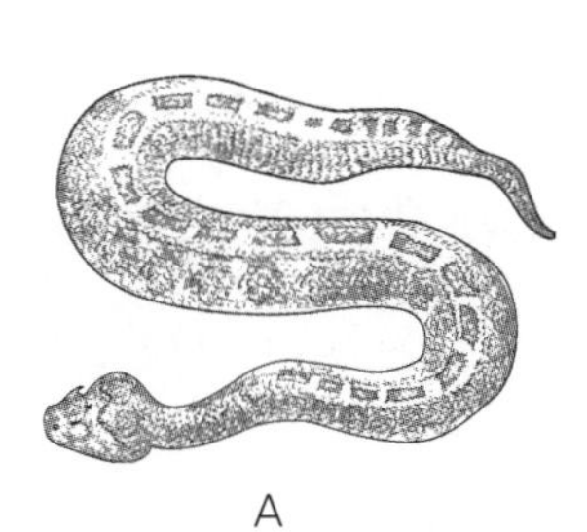

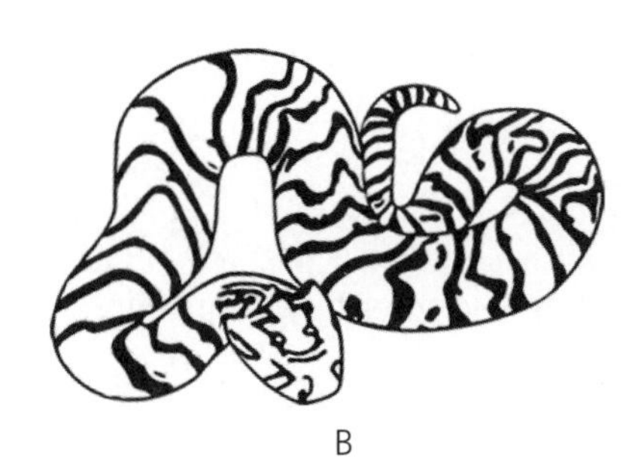

a Use the key on page 208 of the student book to determine which snake is *Bitis arietans*.

b List the features that allowed you to identify it.

....

....

c According to the key, what can you say about the other two snakes?

....

....

2 A key uses a series of questions to distinguish each species from the rest.

Victor uses three questions to distinguish these four cats. His first question separates them into two groups of two. His next questions separates the members of each group.

Put the cats' names into the correct boxes in the table below.

cats with spots		cats with no spots	
cats with spots in groups	cats with separate spots	cats with rounded ears	cats with pointed ears

E

Make a numbered key to identify these six birds.

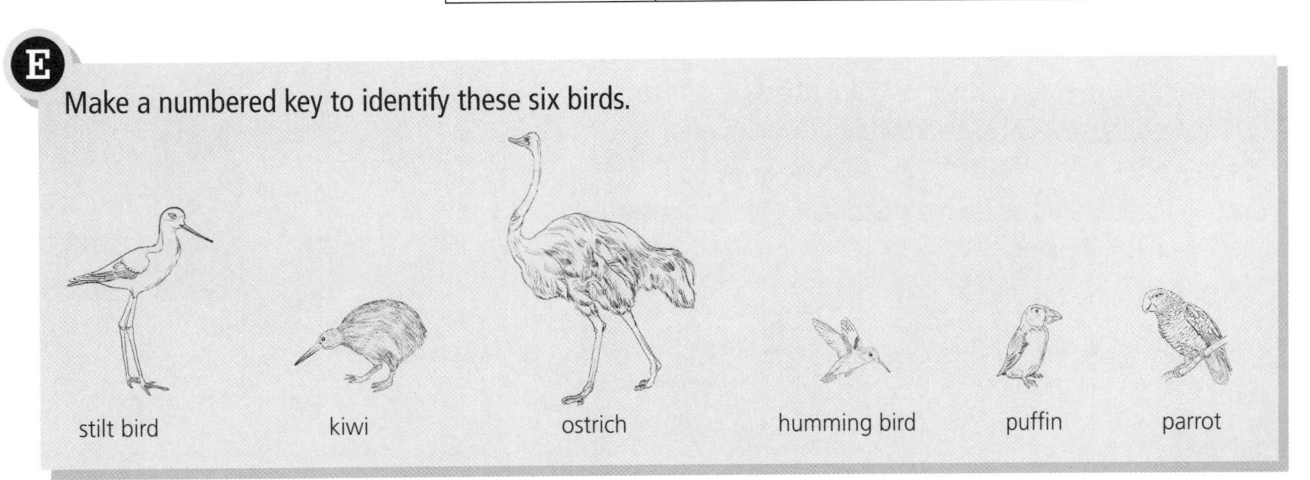

17.2 What makes us different?

1 Complete these sentences about inheritance using words from the box below.

We inherit some characteristics from each .... because we inherit .... from them. Each egg and sperm cell contain .... of each parent's genes – a random selection – so each fertilised egg has a .... combination of genes. These genes .... our growth and development by controlling our .... Some genes are similar in everyone. Others give us unique ....

genes	cells	parent	characteristics	influence	unique	half

2 Label the diagram to show how genes from each parent get into all an embryo's cells.

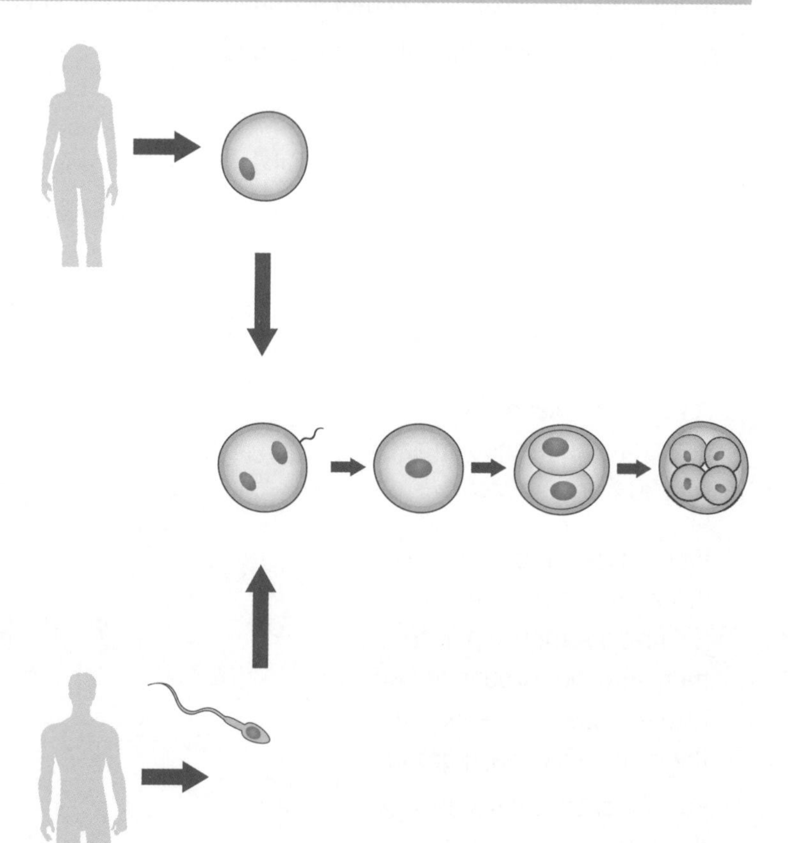

3 Write whether the genes in these groups of cells are the 'same' or 'different'.

a A mother's body cells

....

b The cells in a brother and sister

....

c A mother's egg cells

....

d A father's sperm cells

....

e Each fertilised egg a man and woman produce

....

f The cells in a growing embryo

....

E

Some animals have a gene that makes their brain cells produce less of a certain enzyme. Rats with this gene are more aggressive and antisocial than others. Scientists compared the level of antisocial behaviour among people with and without the gene. Some grew up in happy homes and had no maltreatment. Others suffered severe maltreatment.

a Describe the results.

b Do genes always decide your exact characteristics?

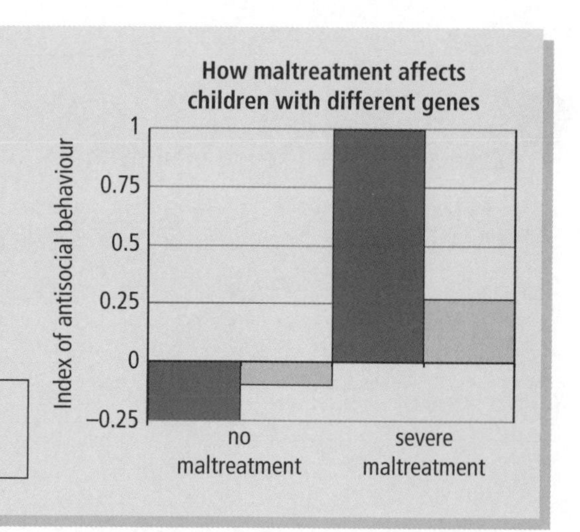

17.3 Chromosomes

1 This diagram shows a human cell.

a Add labels to the diagram to show where the genes are stored in a cell.

b Describe two differences between the diagram and a real cell.

....

....

....

....

c How can you tell whether a cell is from a boy or a girl?

....

....

2 Decide whether these sentences are true for 'genes', 'chromosomes' or 'both'.

a Found in the nucleus ....

b Inherited from our parents ....

c Visible just before a cell divides ....

d Made from giant molecules called DNA ....

e Make from special sections of DNA ....

f There are thousands in each cell ....

g There are 23 pairs in each cell ....

3 Label the diagram on the right to show why fertilised eggs are equally likely to be male or female.

4 The diagram shows the chromosomes from a child with Down's syndrome. Children born with Down's syndrome have a range of physical features. These include wide, flat faces and a short neck. They grow more slowly than normal and tend to be short as adults.

1 2 3 4 5
6 7 8 9 10 11 12
13 14 15 16 17 18
19 20 21 22 X Y

a Is this child a boy or a girl?

....

b What is unusual about the child's chromosomes?

....

c Older mothers are more likely to have children with Down's syndrome. Suggest how the child got this combination of chromosomes.

....

17.4 Investigating inheritance

1 Draw lines to link the correct words from each column to make six sentences about Mendel's work.

To learn about inheritance, Mendel looked at	it was easy to control which male pollen fertilised each female ovule.
He worked with pea plants because	the features pea plants passed to their offspring.
When pollen from tall plants fertilised short pea plants	some of their offspring (the third generation) were short.
When tall plants from the second generation fertilised each other	the dominant 'factor' stops the other one having an effect.
Mendel realised that	all the offspring in the second generation were tall.
When peas inherit two different 'factors'	two 'factors' control each inherited feature – one from each parent.

2 Mendel's 'factors' are now called *genes*, but we still use the code he invented to explain how they are inherited. Decide whether plants with the following genes would be 'tall' or 'short'.

a TT........................

b tt........................

c Tt........................

3 Label the diagram on the left to show why all the second generation are tall.
Complete the diagram on the right to show why some of the third generation are short.

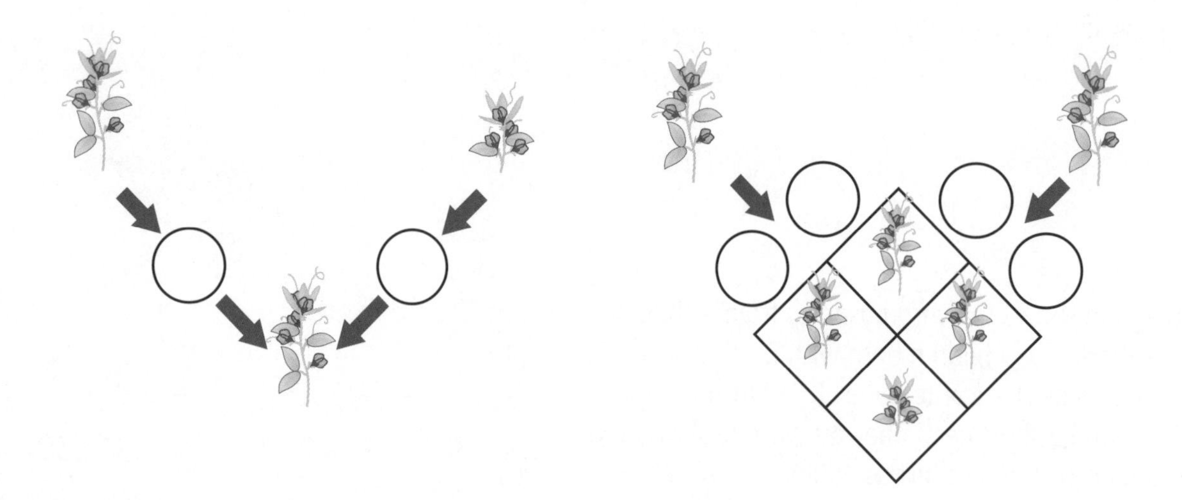

4 Decide whether these genes are 'dominant' or 'recessive'.

a Shaun has red hair. His wife and child have brown hair. The red hair gene is ....

b Seb has two different eye colour genes. His eyes are brown so this gene is ....

c Nadira has two versions of a gene. One causes a disease. The other does not. Nadira does not have the disease, so the gene that causes the disease must be ....

17.5 Selective breeding

1 Draw lines to link words from each column to make five sentences about selective breeding.

Wild cats show a lot of variation	by selecting which cats were allowed to breed together.
Pet cats were produced	so eventually all the offspring had the desirable features.
Each parent cat had some desirable features	because each cat inherits a unique combination of genes.
The best offspring were used to produce the next generation	different breeders produced cat breeds with very different characteristics.
Over many generations	so their offspring could inherit them from both parents.

2 The diagram shows five different cows.

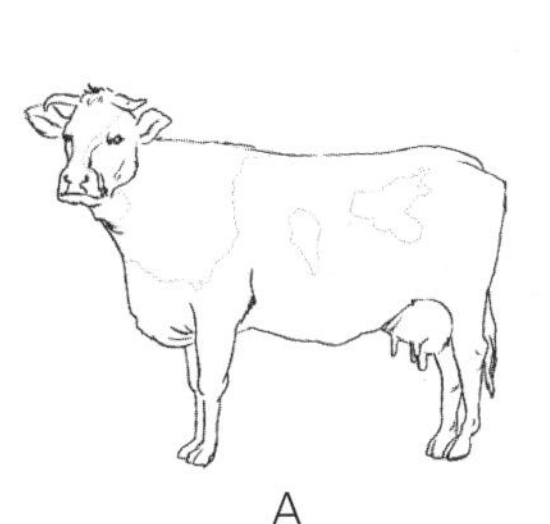

A
high milk yields

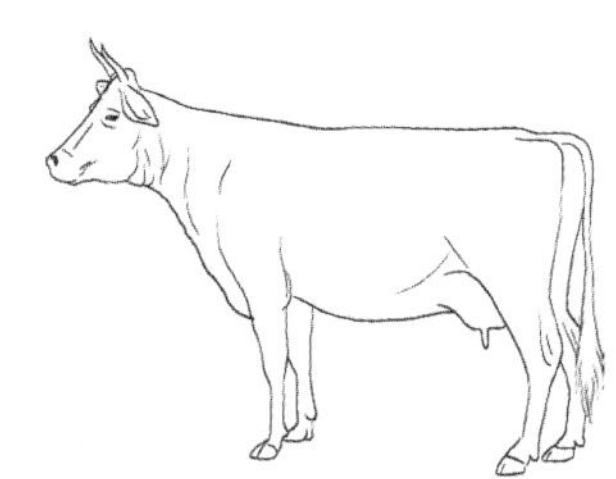

B
strong

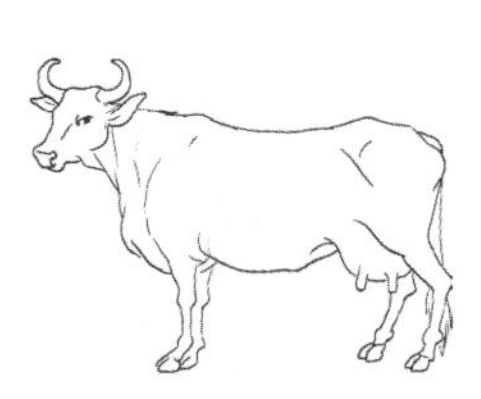

C
resists disease

D
large

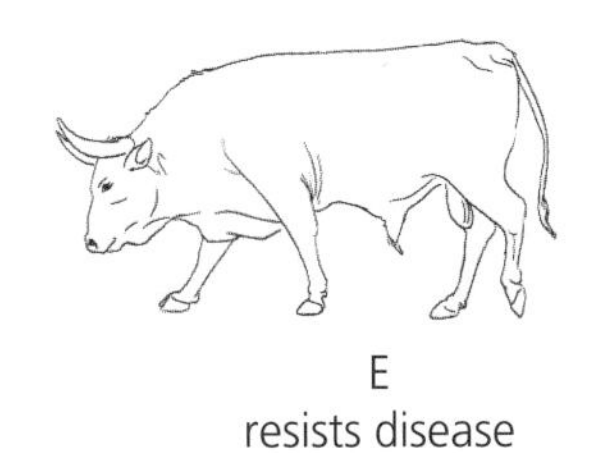

E
resists disease

Choose which cows to breed to produce each of the following:

a A healthy cow for milking ....

b A cow to pull a plough ....

E

Wild bananas are short and full of hard seeds.

Suggest how wild bananas were selectively bred to produce the bananas we eat now.

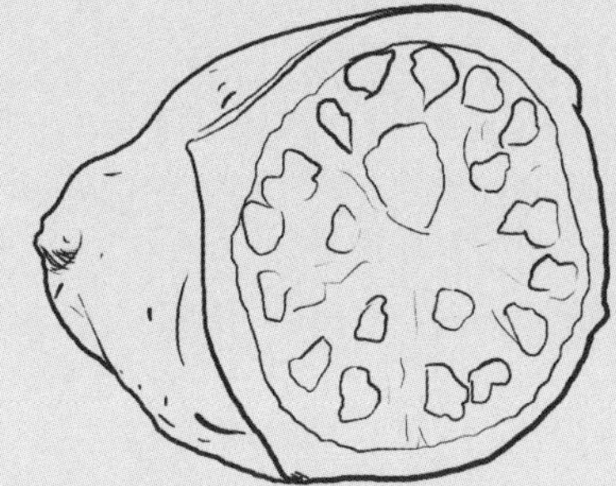

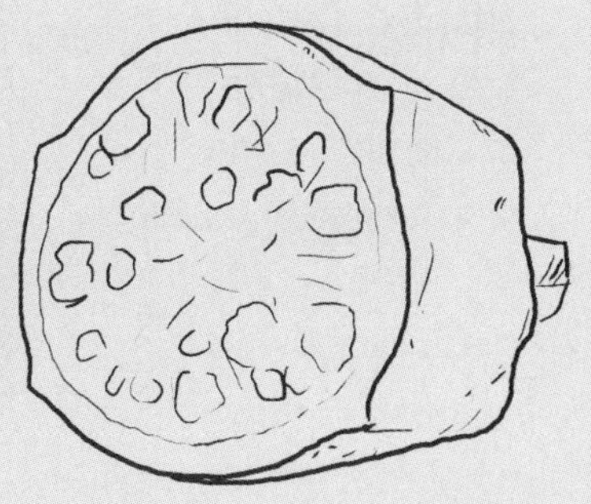

17.6 Developing a theory

1 Read the following paragraph and fill in the gaps with words from the text box below. Each word may be used once, more than once, or not at all.

Theories .... how or why things happen. They usually form gradually as scientists collect observations, ask ...., suggest possible .... and collect .... to support them. Related explanations may eventually link to form a theory. Each .... theory explains a lot of .... and answers a lot of questions.

evidence	observations	questions	explanations	scientific	explain

2 Draw lines to link each observation with the correct explanation.

Observation
Rocks change gradually over many years.
Fossilised mammals have skeletons similar to modern animals but not identical.
Different birds are found on different Galapagos islands.
The birds on each island have beak shapes that suit the food they eat.

Explanation
Birds with well-adapted beaks survive and pass their genes to the next generation.
Mammals have changed over time.
The Earth is very old.
Over many generations, each population changed in different ways.

E

The diagram shows the bones in four mammal's limbs.

a Mammal skeletons provide strong evidence that they share the same ancestor. Explain why.

b Natural selection relies on survival of the fittest or best adapted. Describe two adaptations of the bat's wing which make it useful for flight.

c Describe two features of the whale's fin that make it useful for swimming.

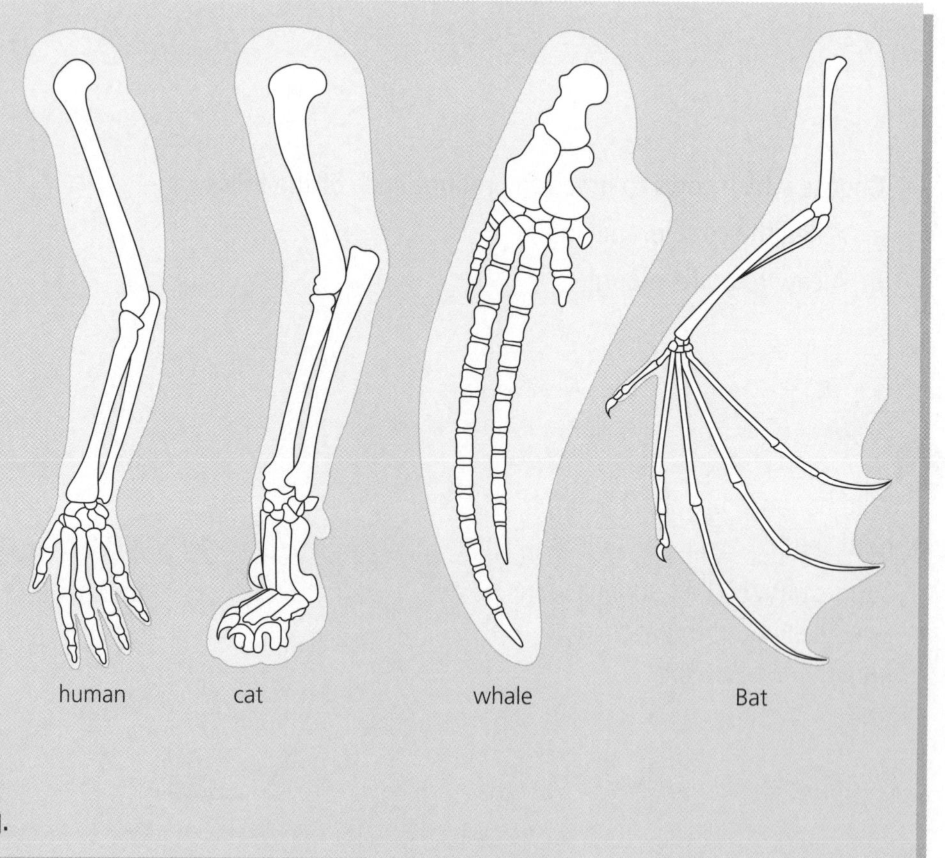

17.7 Darwin's theory of evolution

1 Read the following paragraph and fill in the gaps with words from the text box below.

Members of a species are not all the same. Some have .... that make them more likely to survive and raise .... They pass these useful characteristics to the next .... Over many generations these characteristics become more .... This gradual change is ....

characteristics	common	evolution	offspring	generation

2 Link the key words to the correct definitions.

Key words
Evolution
Variation
Overproduction
Survival of the fittest
Natural selection
Species formation

Definitions
The huge number of offspring that most living things produce.
The way environmental factors like predators and the food supply influence survival.
The way plants and animals change over many generations.
The differences between individuals that make some more likely to survive and reproduce.
The way natural selection can turn populations which become separated into different species.
The survival of individuals best adapted to their environments.

3 Scientists have evidence that polar bears evolved from brown bears.

brown bear polar bear

a Occasionally bears were born with white fur. Suggest how white fur aids survival in the Arctic.

....

b Explain how the genes for white fur could become more common over many generations.

....

....

....

E

Resistance to antibiotics is one of the greatest threats to modern health.

a Do some research. Find out what antibiotics are used for.

b Explain how resistance to antibiotics could evolve.

17.8 Moving genes

1 Complete these sentences about genetic engineering using words from the text box.

Living things can be genetically .... to make useful .... The human gene for insulin has been transferred to bacteria, plants and animals. These now produce .... for people with .... who cannot make their own. The plants produce insulin in their seeds and the animals produce it in their .... The insulin has to be .... from the bacteria, seeds or milk before it can be given to .... as an injection.

engineered insulin milk diabetics separated products diabetes

2 Add the correct word to finish each of these sentences.

a Adding genes to cells is ....

b The small loops of DNA in bacteria are ....

c Using genetically engineered animals to produce medicines is ....

3 Draw lines to match each observation to the correct explanation.

Observation
Scientists can move genes from one species to another
Any gene put into a fertilised egg is copied every time it divides
It is easier to transfer genes to bacteria than to plant or animal cells
New genes are often added to plasmids
Bacteria with added genes can make large amounts of useful proteins

Explanation
because bacteria have no nucleus. Their genes are loose in the cytoplasm.
because these can be moved in and out of bacterial cells.
because the genes in all living things are made from the same chemicals.
because bacteria can grow and reproduce rapidly in large tanks.
so every cell in the plant or animal produced has a copy of the gene.

4 Human insulin was the first product made by genetic-engineered bacteria.

Do some research. Find two other medical products made this way and describe what they are used for.

....

....

....

....

....

....

human insulin

17.9 Using genes

1 Complete these sentences about artemisinin using the words in the text box.

Artemisinin is the best cure for .... Unfortunately it is expensive to extract from plants and too .... to make in a laboratory. Some scientists are trying to make artemisinin cheaper by .... plants that produce more of it. Other scientists have used .... engineering to put artemisinin genes into bacteria and .... cells. Yeast cells grow and .... rapidly so they can make large quantities of the ....

genetic   malaria   yeast   reproduce   breeding   medicine   complicated

2 The graph shows how much artemisinin different batches of plants produced.

a Label the most common yield.

b Label the plants you would use to begin selective breeding.

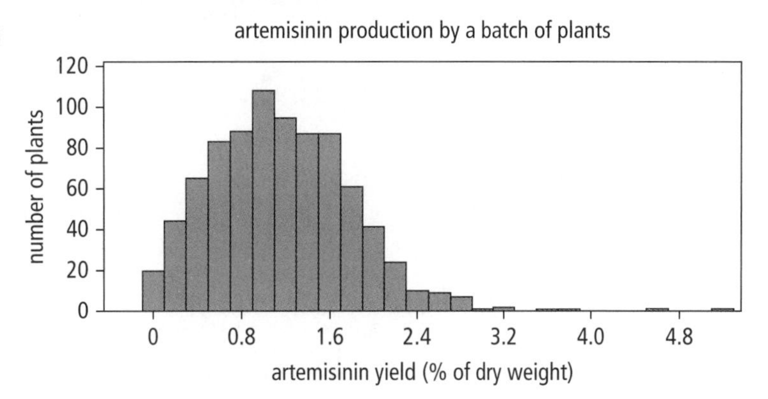

3 The diagram shows how the artemisinin gene can be added to bacteria.

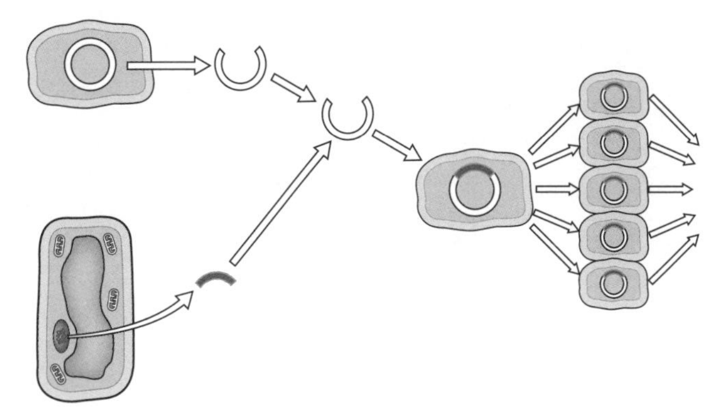

Add letters a-f to the correct parts of the diagram to show where the following labels should go.

a Genes for the enzymes that make artemisinin are taken from artemisia plants.

b Bacterial plasmids are removed and cut open.

c Genes from artemisia plants are added to the bacterial DNA.

d The bacterial plasmid is put back into the bacteria.

e Bacteria copy their genes and reproduce rapidly.

f All the new bacteria make artemisinin.

4 Making yeast cells produce artemisinin is a much more complicated process than making bacteria produce insulin. It is an example of *synthetic biology*. Do some research. Find out what synthetic biology involves and what it might be used for in the future.

Checkpoint-style practice questions

1 The diagram shows the main organs in the human body.

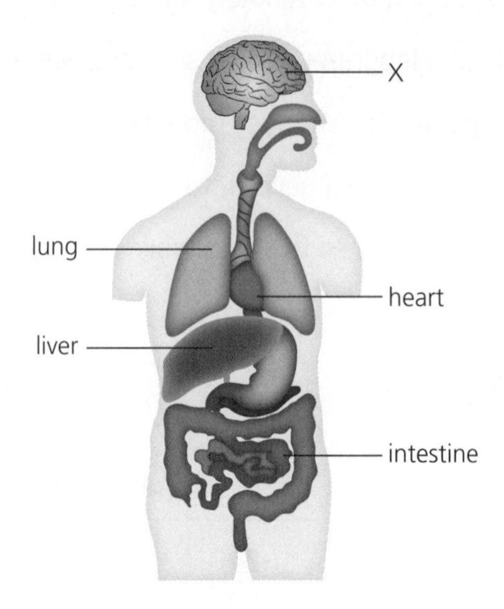

a Which organ takes in glucose?

.... [1]

b Which organ system are the lungs part of?

.... [1]

c Describe the role of the lungs?

.... [1]

d Which organ system is the heart part of?

.... [1]

e Name the main type of tissue in the heart?

.... [1]

f Name the main type of tissue in organ X

.... [1]

2 The diagram shows part of a human shoulder joint.

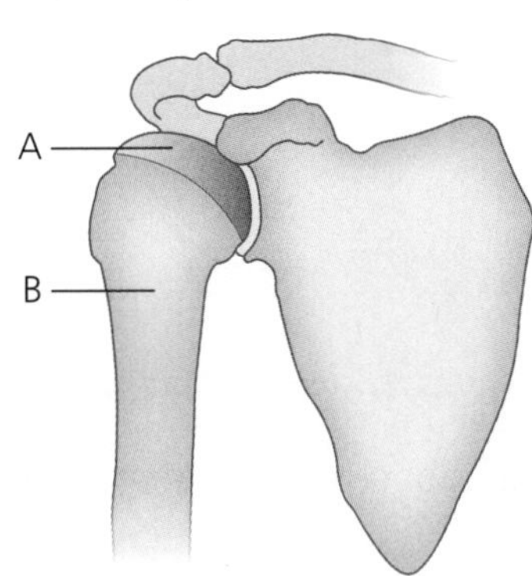

a i Name components A and B.

.... [1]

ii Why does our skeleton have joints?

.... [1]

iii What type of joint is this?

.... [1]

b Several organ systems are involved in producing arm movements. Fill in the table below to show their functions. [3]

Organ system	**Function**
Skeletal system	
Muscular system	
Nervous system	

c The circulatory system brings extra blood to working muscles.

i Name one substance muscles remove from blood.

.... [1]

ii Name one substance muscles add to blood.

.... [1]

3 Kassa is investigating how bacteria affect the pH of milk. He adds different numbers of bacteria to six samples of pasteurised milk.

a The table shows the results.

Number of bacteria added (billions)	**pH of milk after 2 days**
0	6.0
2	5.7
4	5.1
6	4.8
8	4.2

On graph paper, plot a line graph using these results and label the axes

[3]

b Describe the relationship shown on the graph.

....

.... [1]

c What pH would you expect if 5 billion bacteria had been added?

.... [1]

d Suggest why Kassa used pasteurised milk for the experiment.

....

.... [1]

e All investigations involve several variables.

i Name the variable Kassa changed.

.... [1]

ii Which variable was measured?

.... [1]

iii Suggest two variables Kassa should have controlled.

1 ....

2 .... [2]

f Micro-organisms like bacteria can be useful.

i Name two foods made using bacteria or yeast.

1 ....

2 .... [2]

ii Describe one other way in which microbes are useful.

.... [1]

4 The animal below is a river otter. It eats fish.

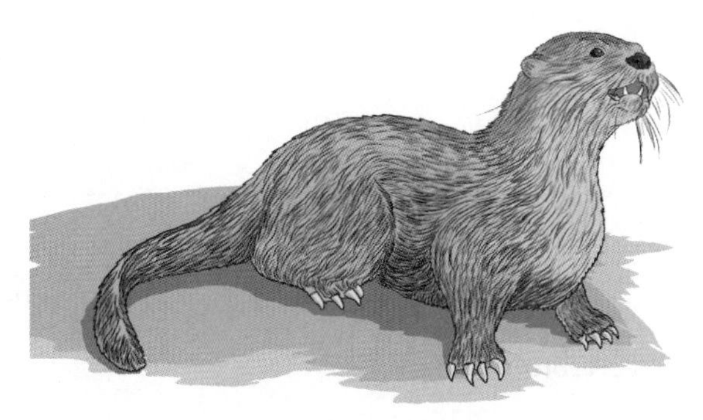

a Suggest two adaptations that would help these otters to survive.

1 ....

2 .... [2]

b Use the information below to write a food chain.

Otters eat fish. The fish feed on small invertebrates called zooplankton. Zooplankton feed on tiny phytoplankton, which can make their own food. Many otters are eaten by alligators. [1]

c Alligators are carnivores. What does this mean?

.... [1]

d Which organism in the food chain is a producer?

.... [1]

e Name a predator from the food chain and its prey.

predator ....

prey .... [2]

f The number of fish in the otters' rivers are falling.
Explain how this could affect the number of otters.

....

.... [1]

5 This iguana has just laid her eggs in a burrow.

a Name the classification group the iguana belongs to.

.... [1]

b List two features that every animal in this group has.

1 ....

2 .... [2]

c What feature does the iguana share with every other vertebrate?

.... [1]

d The chart shows the tail lengths of a group of iguanas.

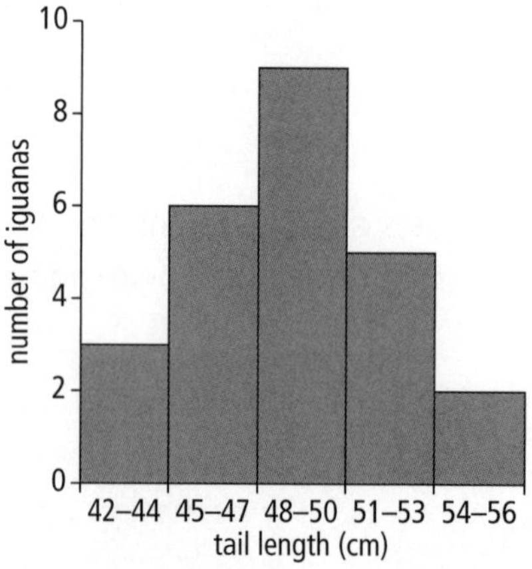

number of iguanas
10
8
6
4
2
0
42–44 45–47 48–50 51–53 54–56
tail length (cm)

i How many tails were measured?

.... [1]

ii Do their tail lengths show continuous or discontinuous variation?

.... [1]

6 Plant and animal cells form tissues, organs and organ systems.

a Each word in the list below is an organ system, organ or tissue.

fatty circulatory stomach connective brain respiratory

Write each word in the correct column of the table.

organ systems	organs	tissues

[3]

b The cells in each tissue have different functions.
Describe the main function of each of these plant cells.

i root hair cell

.... [1]

ii leaf cell

.... [1]

c The diagram shows a leaf cell.

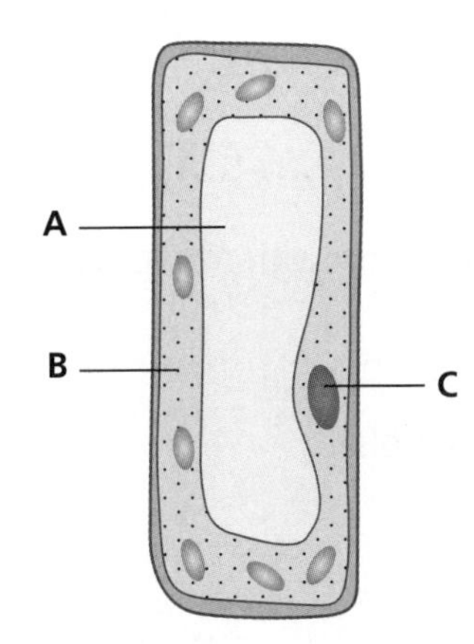

Name parts A, B and C.

A .... [1]

B .... [1]

C .... [1]

d Name one part of a leaf cell which is not found in root hair cells.

.... [1]

7 The diagram shows a leaf in sunlight.

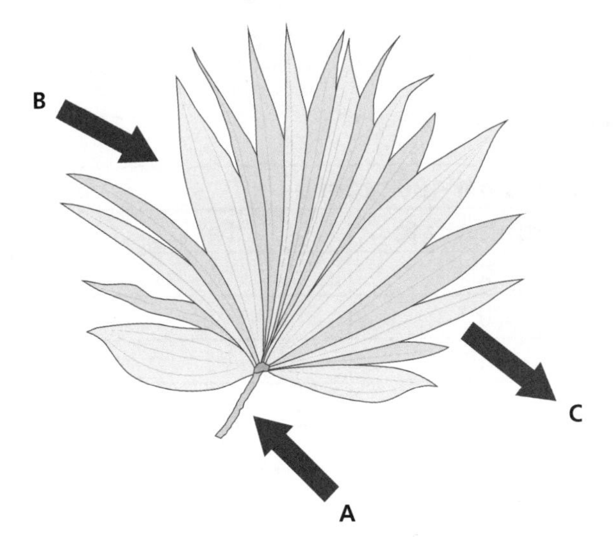

a Substances are moving in and out of the leaf.
Name substances A, B and C.

A .... [1]

B .... [1]

C .... [1]

b Several types of cell are involved in the transport of water.

i Name the cells which absorb water.

.... [1]

ii Which cells form the tubes which carry water to leaves?

.... [1]

iii Describe how most water leaves a plant.

....

.... [2]

c Water carries minerals to plant cells. The table shows what some of these are used for.

Mineral	Use
nitrogen	to make protein
phosphorus	to transfer energy
potassium	to make fruit
magnesium	to make chlorophyll

A plant receives plenty of light but its leaves have turned yellow. Which mineral is it missing?

.... [1]

8 The table shows the mass of each nutrient in foods A-E.

Food	Mass (grams per 100 grams)		
	Protein	Carbohydrate	Fat
A	5	30	0
B	30	0	20
C	3	30	25
D	0	10	0
E	12	2	12

a Name two nutrients not shown in the table.

.... [2]

b Different nutrients have different roles in the body.

i Which food is best for growth and repair?

.... [1]

ii Which food supplies most energy?

.... [1]

iii Which food could be meat?

.... [1]

c Name one component of food which is not digested and write why it is an important part of the diet.

....

.... [1]

d Calculate the mass of carbohydrate in 50 g of food A.

.... [1]

e Name two different types of carbohydrate.

.... [1]

f The large molecules in food are broken down during digestion. Name the proteins used to speed up this process.

.... [1]

9 Fabio puts 1 cm^3 of jelly into three test tubes. The jelly is made of protein. He adds equal amounts of enzyme to each tube and leaves them for two hours.

Test tube	Enzyme added	Temperature of test tube (°C)	Observations after 2 hours
A	enzyme from stomach	37	All broken down
B	enzyme from stomach	100	Unchanged
C	enzyme from saliva	37	Unchanged

a Fabio compares tubes A and C. What conclusion should he write?

....

.... [1]

b What can Fabio conclude from tubes A and B?

....

.... [1]

c In a second investigation Fabio adds stomach enzyme to the two samples of jelly shown below. He keeps them both at 37°C. The single cube takes 90 minutes to break down.

i Predict whether the smaller pieces will take more or less time.

.... [1]

ii Give a reason for your answer to part i.

....

.... [1]

d What is produced when proteins are broken down?

.... [1]

10 A sports scientist measured four students' resting heart rates. They each ran for ten minutes. Then the scientists measured how long their heart rates took to return to normal.

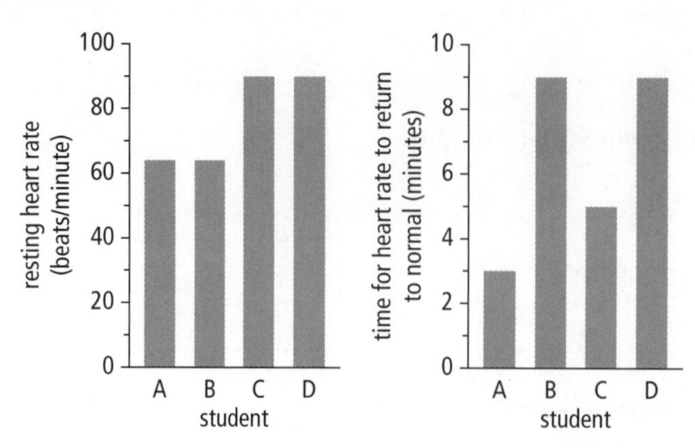

a Which two variables were controlled during the investigation?

1 ....

2 .... [2]

b Every student's heart rate increased when they ran.
Explain why they needed to pump more blood through their muscles.

....

.... [2]

c Blood contains two types of cell. Describe what the red blood cells do.

....

.... [2]

d Which two students had the lowest resting heart rates?

.... [1]

e Which two students took longest to recover?

.... [1]

f Use the information in the bar charts to decide which student was fittest.

.... [1]

11 Tomas used a microscope to look at blood vessels. He measured the thickness of their walls. The table shows his findings.

Vessel	Thickness of wall (mm)
A	0.60
B	0.35
C	0.06

a Suggest which vessel carries blood from the heart under high pressure.

.... [1]

b Vessel C is from a muscle. It has very thin walls that dissolved gases can move through. Explain why this is useful.

....

.... [1]

c These two arteries are from different people.

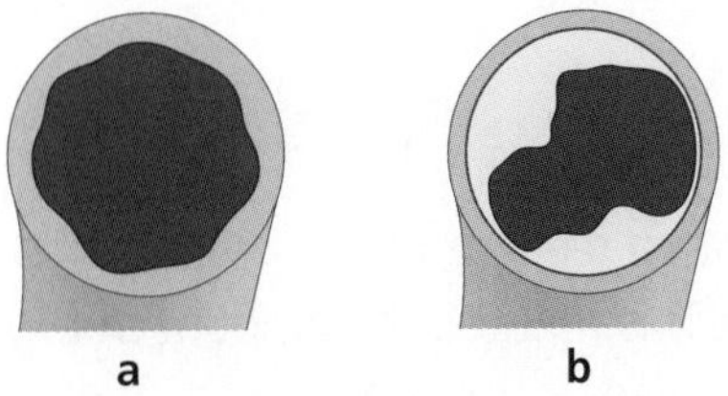

i Which would let blood through more quickly?

.... [1]

a b

ii Which is most likely to be from a patient with high blood pressure?

.... [1]

iii Give a reason why having a high blood pressure is dangerous.

.... [1]

d A good diet keeps your circulatory system healthy.

Choose two items from the list that doctors recommend you to eat.

saturated fatty acids	**salt**	**fish oils**
unsaturated fatty acids	**sugar**	**nuts and seeds**

1 ....

2 .... [2]

12 The diagram shows a fetus in the uterus.

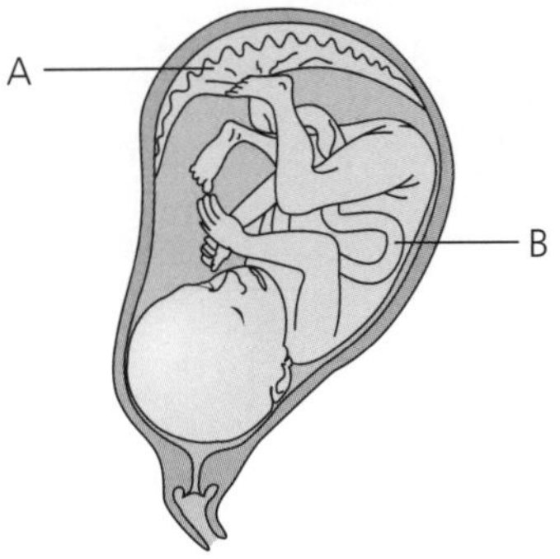

a Name parts A and B.

A

B

A .... [1]

B .... [1]

b In the placenta, blood vessels from the fetus are surrounded by its mother's blood.

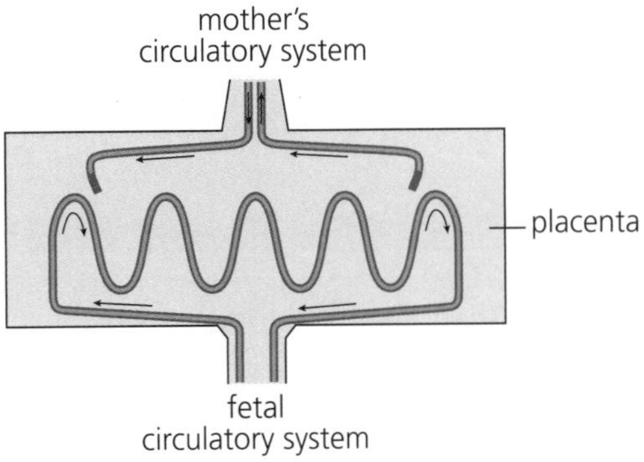

mother's circulatory system

placenta

fetal circulatory system

Write the following letters in order to describe the route oxygen takes to get to muscle cells in the fetus.

A mother's red blood cells

B fetal red cells in umbilical cord

C mother's alveoli

D fetal muscle cells

E fetal red cells in placenta

F mother's trachea

.... [3]

c Smokers often feel breathless because cigarette smoke contains carbon monoxide. This reduces the amount of oxygen their red blood cells can carry.

Suggest why the babies of mothers who smoke have a lower birth mass than the babies of non-smokers.

.... [1]

d Pregnant women are advised not to drink alcohol. Explain why.

....

.... [2]

13 Bilharzia is a tropical desease. Scientists compared the average body mass of boys from areas where bilharzia is common (solid line) with those from areas free from the disease (dotted line).

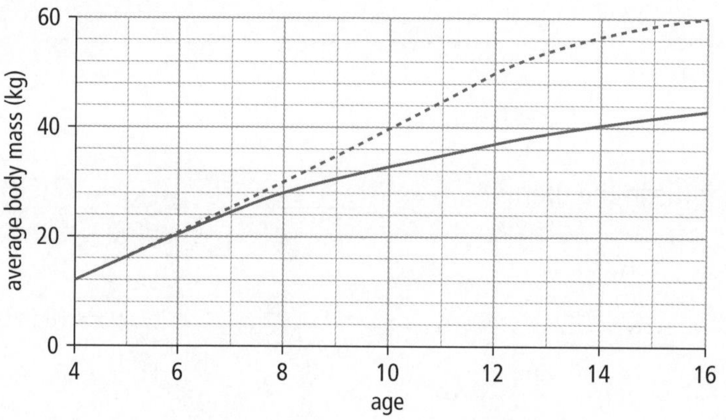

a At what age does their average body mass start to be different?

.... [1]

b Calculate the size of this difference in body mass when the boys in both groups are 14.

.... [2]

c The worms that cause bilharzia reproduce in their victim's blood vessels. Suggest how they could reduce a boy's growth.

.... [1]

d Suggest extra data the scientists could collect to extend their investigation.

.... [1]

14 Freya wants to find out how the rate of photosynthesis in pondweed is affected by light intensity. She sets up this apparatus.

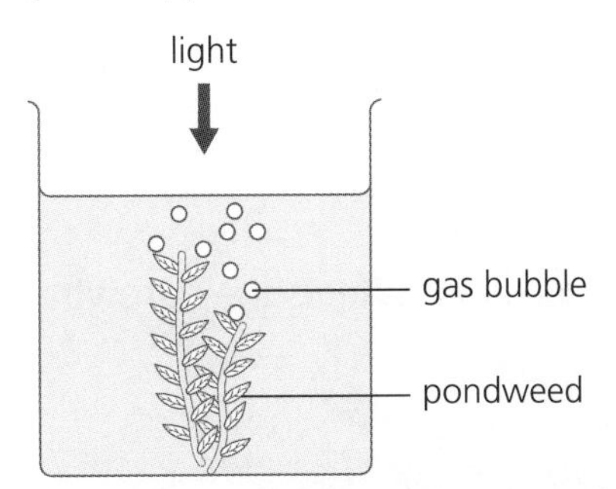

Freya moves the pondweed different distances from the lamp. At each distance she counts the bubbles given off per minute.

Distance (cm)	Number of bubbles per minute
10	30
20	16
30	10
40	6
50	4
60	3

a On graph paper, draw a graph to display these results. [3]

b Name the reaction that is producing the gas in the bubbles. .... [1]

c Write a word equation for the reaction named in part b.

.... [1]

d Explain why the number of bubbles per minute drops when the lamp is further away.

....

.... [1]

e How could Freya check that her results were reliable?

.... [1]

f In a second investigation, Freya adds sodium hydrogen carbonate to the water around the pondweed. This increases the concentration of carbon dioxide in the water.

Predict how this affects the number of bubbles given off per minute.

.... [1]

g Counting bubbles is not very accurate. Suggest how Freya's method could be improved.

....

.... [1]

15 Sara investigates how the distance between seedlings affects their growth.

Distance between seedlings (cm)	Average height of seedlings (cm)
1	3
5	7
10	12

a List two variables that Sara should have controlled.

1 ....

2 .... [2]

b Do Sara's results support the idea that seed dispersal is important?

.... [1]

c Explain why seedlings grow better if they are further apart.

....

.... [2]

d Suggest how these seeds are dispersed.

i They are very light and covered with tiny hairs.

.... [1]

ii They have juicy fruits which turn bright red when they ripen.

.... [1]

iii Their fruits are covered in tiny hooks.

.... [1]

e Scientists often want to preserve seeds. How could they prevent them from germinating?

.... [1]

16 The diagram below shows the energy transfers in a food chain.

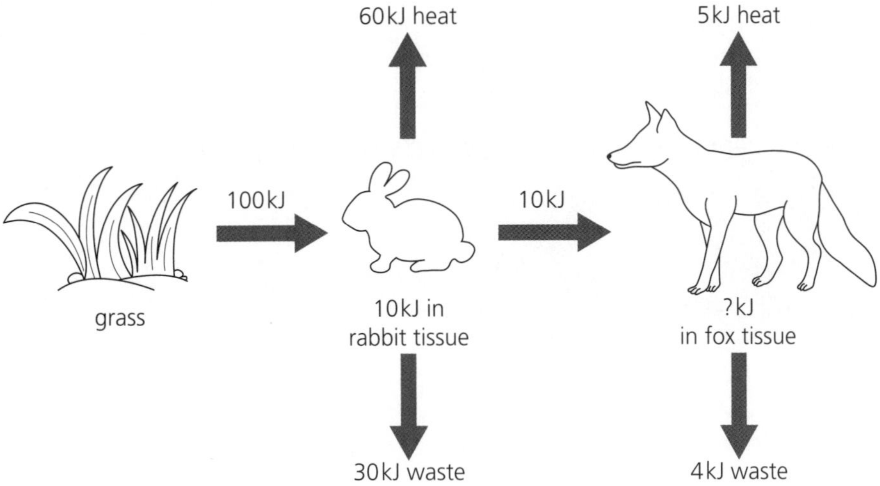

The rabbit uses about 10% of the energy in its food to build new tissues. When the rabbit is eaten this energy is transferred to a fox.

a The numbers on the diagram show what happens to each 100 kJ of energy. Calculate how much energy the fox uses to build new tissues.

.... [1]

b List two ways energy is lost from the food chain.

1 ....

2 .... [2]

c In the space below, draw a pyramid of biomass for this food chain. [2]

17 Tenisha investigates decay. She takes two slices of bread from the same loaf. She adds 10 cm^3 of water to each slice. Then she puts one slice in the fridge and one slice in a warm place.

After 3 days she compares the slices.

Bread	Observations
Kept in fridge	Green in one corner
Kept warm	Green all over

a What type of organism made the damp bread turn green?

.... [1]

b Which variable did Tenisha change in her investigation?

.... [1]

c What can you conclude from her results?

.... [1]

d Explain why decay is important in food webs.

.... [1]

e The graph shows how the number of bacteria changed when some sewage leaked into a pond.

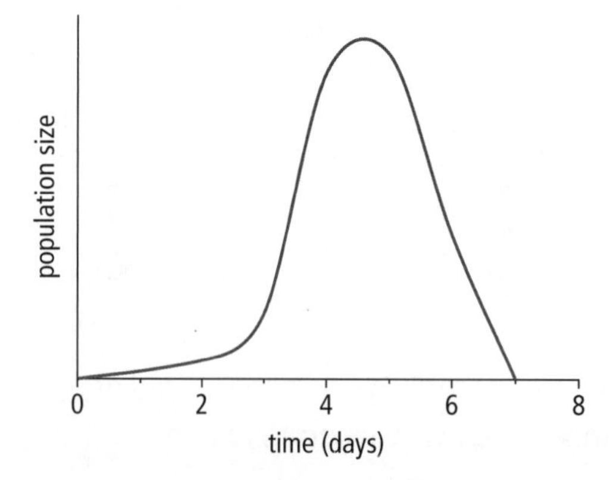

i On which day was the population increasing most rapidly?

.... [1]

ii When did the bacteria begin to die faster than they reproduced?

.... [1]

18 A student left two flasks out on a sunny day.

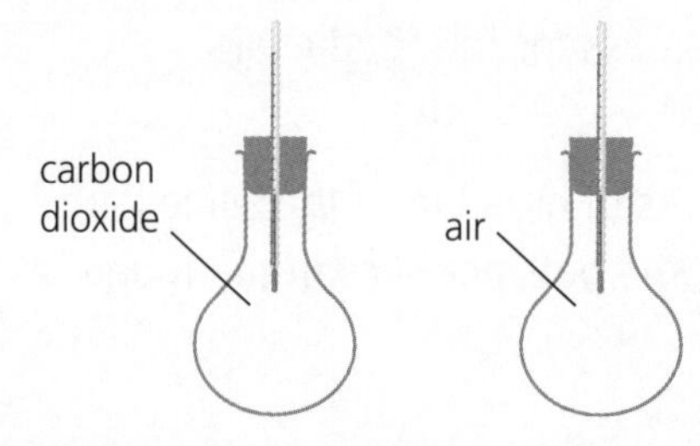

At the end of the day she plotted graphs to show how the temperature inside each flask had changed.

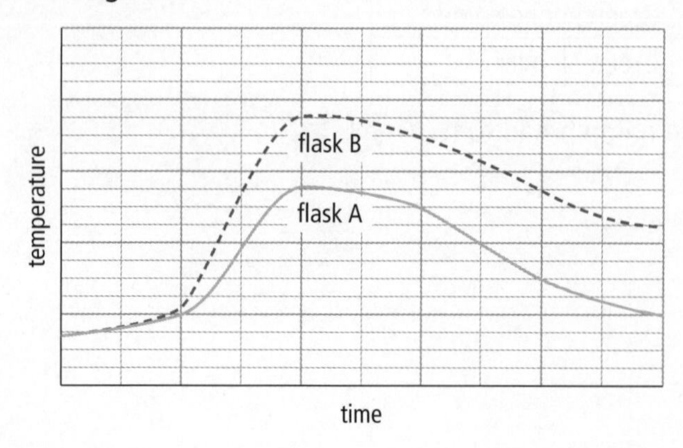

a Which flask contained carbon dioxide?

.... [1]

b Give two reasons why the amount of carbon dioxide in the atmosphere is rising?

1 ....

2 .... [2]

c Describe one environmental problem that adding carbon dioxide to the atmosphere could cause.

....

.... [2]

d Name another environmental problem caused by pollutants in the atmosphere.

.... [1]

19 Modern maize plants were produced by selective breeding. They produce much bigger cobs than ancient varieties did.

ancient variety

modern variety

a Write the following letters in order to describe how modern maize plants were produced.

A grow the seeds into new plants

B select the plants with the biggest cobs

C repeat this for many generations

D breed these together to produce new seeds

.... [2]

b Suggest one other characteristic that is useful in a food crop.

.... [1]

c Explain why each new plant inherits characteristics from each of its parents.

....

.... [1]

d Individual plants are all slightly different. What name is given to their differences?

.... [1]

e Darwin realised that living things could also be changed by 'natural selection'.
Write the following letters in order to describe how evolution takes place.

A the next generation inherit their useful characteristics

B members of a species show a lot of variation

C over many generations these characteristics become more common

D those best adapted to their environment survive and reproduce

.... [2]

Cambridge IGCSE® practice questions

Are you ready for the step up to Cambridge IGCSE? Here are some questions from real Cambridge IGCSE exam papers. Use your knowledge from Cambridge Secondary 1 Biology to answer these questions.

1 Fig. 1.1 shows the bones and muscles of a human leg.

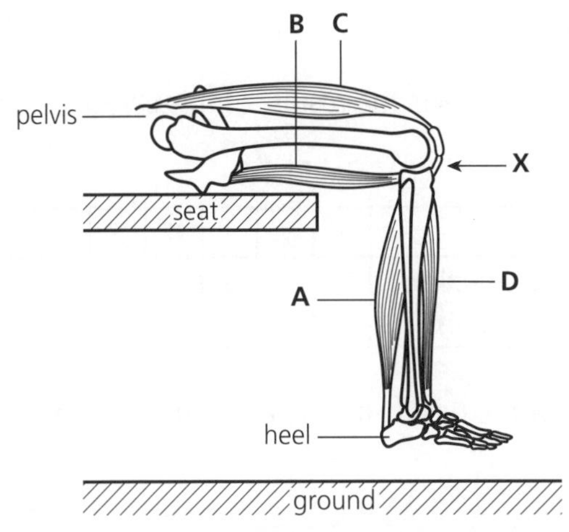

Fig. 1.1

Muscles in the leg work antagonistically.

a State which muscle is antagonistic to muscle **A**.

.... [1]

b Explain what is meant by *antagonistic*.

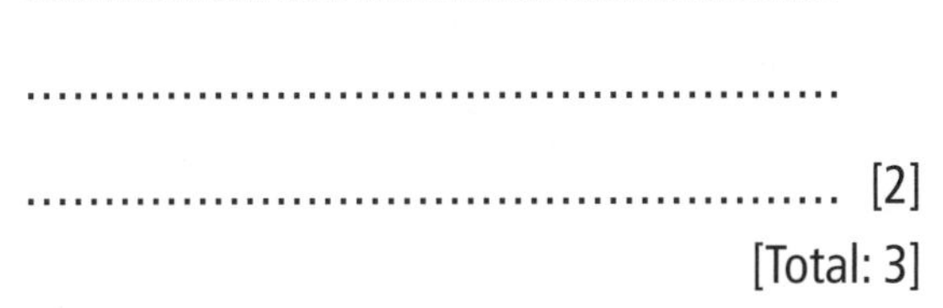

....

....

.... [2]

[Total: 3]

Cambridge IGCSE Biology 0610 Paper 2 Q8 November 2008

2 a Fig. 2.1 shows the variation in the height of human adults in an African population.

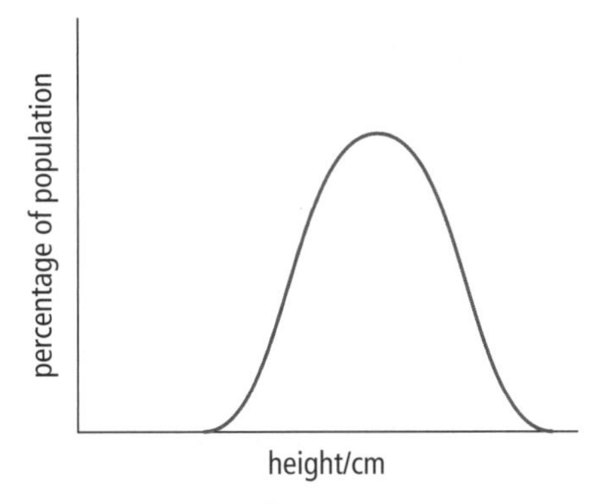

Fig. 2.1

State the type of variation shown by this data.

.... [1]

b In Britain 42% of the population have blood group A. The frequency of the other blood groups is: B (9%), AB (3%) and O (46%).

i Plot the data, as a bar chart, on Fig. 2.2. [2]

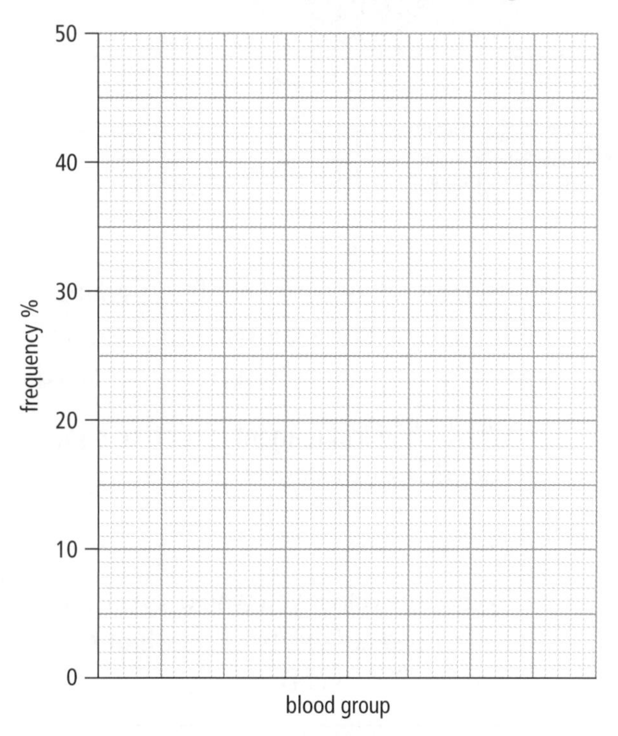

Fig. 2.2

ii Complete the following sentence.

Height is controlled by environment and by genes but human blood groups are controlled only by

.... [1]

[Total: 4]

Cambridge IGCSE Biology 0610 Paper 2 Q3 June 2007 – not all parts of the question have been used

3 a i Name the two raw materials needed by plants for photosynthesis.

1 ....

2 .... [2]

ii Name the gas produced by photosynthesis.

.... [1]

b Fig. 3.1 shows a leaf, with white and green regions, that is attached to a plant. The plant had been kept in the dark for 48 hours and then a lightproof, black paper cover was placed over part of the leaf.

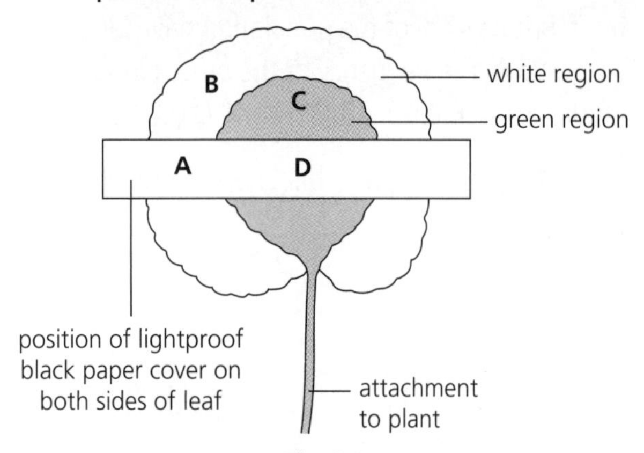

Fig. 3.1

The plant is left under a light for 24 hours. After this time the leaf is removed from the plant and is tested for the presence of starch.

i Which chemical reagent is used to show the presence of starch?

.... [1]

ii Record the colour you would see, if you had carried out this test, in each of the areas **A**, **B**, **C**, and **D**.

Area	Colour
A	
B	
C	
D	

[4]

iii Explain the results for each of the following areas.

area **B** ....

....

area **D** ....

.... [2]

[Total: 10]

Cambridge IGCSE Biology 0610 Paper 2 Q4 June 2008

4 A student investigated the effect of changing pH on the rate of reaction of a digestive enzyme.

a Define the term *enzyme*.

....

....

.... [2]

Table 4.1 shows the results of this investigation.

Table 4.1

pH	1	2	3	4	5	6	7
rate of reaction / arbitrary units	10	15	9	6	3	1	0

b Plot the results as a line graph on Fig. 4.1. [3]

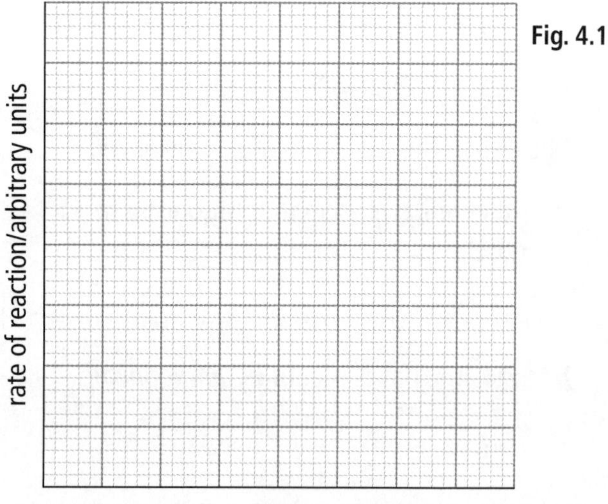

Fig. 4.1

c Suggest where in the human digestive system this enzyme would have been most active.

.... [1]

d The investigation at pH 3 was repeated but the enzyme was boiled before its use. Suggest how and why the results would have been different.

....

....

.... [2]

[Total: 8]

Cambridge IGCSE Biology 0610 Paper 2 Q7 June 2007

5 a Table 5.1 shows the percentage of haemoglobin that is inactivated by carbon monoxide present in the blood of taxi drivers in a city.

Table 5.1

City taxi drivers		Percentage of haemoglobin inactivated by carbon monoxide
day-time drivers	smokers	5.7
	non-smokers	2.3
night-time drivers	smokers	4.4
	non-smokers	1.0

i The carbon monoxide in the blood of these taxi drivers comes from two sources. One source is from vehicle exhaust fumes.

Name the other source of carbon monoxide that may be inhaled by drivers.

.... [1]

ii Using data from Table 5.1, suggest which of these two sources contributes most to the inactivation of the haemoglobin.

Explain your choice.

source ....

explanation ....

....

....

.... [3]

iii Calculate the difference in the percentage of haemoglobin inactivated by carbon monoxide in day-and night-time taxi drivers and suggest a reason for the difference.

difference ....

reason ....

.... [2]

b i Name two other harmful components of cigarette smoke, apart from carbon monoxide.

For each, describe an effect it can have on the body of a person who smokes.

1. component ....

effect ....

....

2. component ....

effect ....

.... [4]

ii Suggest a possible effect that might happen to the fetus of a pregnant woman who smokes.

.... [1]

[Total: 11]

Cambridge IGCSE Biology 0610 Paper 2 Q2 November 2008

6 a Using straight lines, match the names of the flower parts with their functions. One has been completed for you.

anther	allows the passage of the pollen tube to the ovary
petal ———	attracts insects for pollination
sepal	produces pollen grains
style	protects the flower when in bud
stigma	the surface on which the pollen lands during pollination

[4]

b Describe how the stigmas of wind-pollinated flowers differ from the stigmas of insect-pollinated flowers. Relate these differences to the use of wind as the pollinating agent.

....

....

....

....

.... [3]

[Total: 7]

Cambridge IGCSE Biology 0610 Paper 3 Q1 June 2008 – not all parts of the question have been used

7 In Africa, mammals called jackals are quite common. They feed on small herbivores such as young springboks and dik-diks, hunting in packs to catch their prey. They will also eat larger herbivores such as kudu that have been killed by larger predators such as lions.

A farmer in South Africa found that a number of his sheep, while feeding on grassland, were being killed by jackals. He noted that jackals always kill sheep by attacking their necks. He designed a plastic collar for the sheep that covered their necks. None of his sheep have been killed since fitting these collars. Other farmers are now buying the collars to protect their sheep from jackal attack.

a The prey species of the jackal are usually primary consumers.

State the type of food that all primary consumers eat.

.... [1]

b Name the two carnivores identified in the text.

1 ....

2 .... [1]

c Construct a food chain for the jackal to show its relationship with sheep.

.... [2]

d Suggest a reason why jackals survive better when they hunt in packs.

....

.... [1]

e When the farmer started to use collars on his sheep, although none of his sheep were being killed, the population of jackals did not decrease.

Suggest why the number of jackals did not decrease.

....

.... [1]

f Name two structures, found in the neck of a sheep, that could be damaged when jackals attack it.

1 ....

2 .... [2]

[Total: 8]

Cambridge IGCSE Biology 0610 Paper 3 Q1 June 2004 – not all parts of the question have been used

8 The wild dog is one of the smaller African carnivorous mammals. It has disappeared from 25 of the 39 countries where it used to live. Wild dogs hunt in packs, feeding on antelopes, which are grass-eating mammals.

A conservation programme has been started to increase the wild dog population in South Africa. Farmers are worried about numbers getting out of control because wild dogs breed at a very fast rate. However, conservationists are not concerned because the lion is a natural predator of the dogs.

a Wild dogs are carnivorous mammals.

i Define the term *carnivore*.

.... [1]

ii State **one** external feature which distinguishes mammals from other vertebrates.

.... [1]

b i Suggest two reasons why numbers of African wild dogs are decreasing.

1 ....

2 .... [2]

ii Suggest what could happen to the species if numbers continue to decrease.

....

.... [1]

c Using the information in the passage above, construct a food chain for a wild dog, including its predator.

Label each organism with its trophic level.

[4]

d It is important that the wild dog species is conserved.

i Explain the meaning of the term *conservation*.

....

....

.... [2]

ii Outline the measures that could be taken to conserve a mammal, such as the wild dog.

....

....

....

....

....

[3]

[Total: 14]

Cambridge IGCSE Biology 0610 Paper 3 Q2 June 2008 – not all parts of the question have been used

9 Fig.9.1 shows six arthropods, each of which could carry disease organisms.

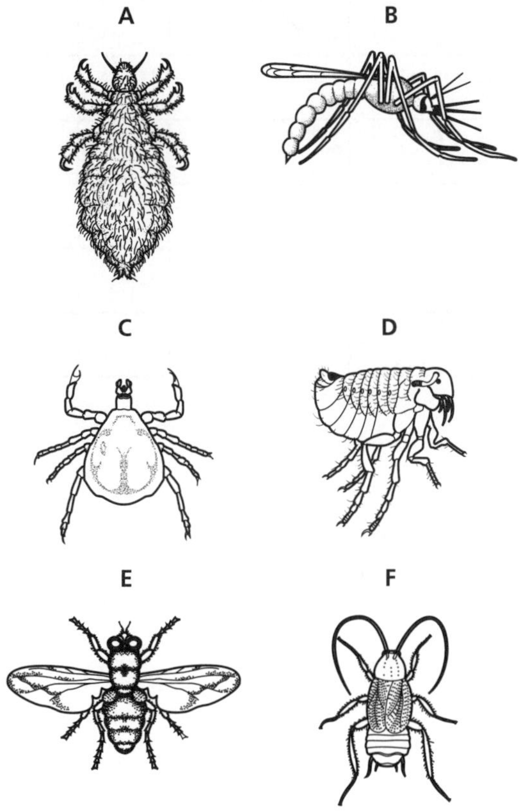

A B C D E F

Fig. 9.1

Use the key to identify each of the arthropods. Write the name of each arthropod in the correct box of Table 9.1. As you work through the key, tick (✓) the boxes in Table 9.1 to show how you identified each arthropod.

Arthropod **A** has been completed for you as an example.

Key

	arthropod
1 a Wings present	go to 2
b Wings absent	go to 4
2 a Wings shorter than abdomen	go to 3
b Wings longer than abdomen	*Musca*
3 a Abdomen long and narrow	*Anopheles*
b Abdomen short and broad	*Periplaneta*
4 a Has three pairs of legs	go to 5
b Has four pairs of legs	*Ornithodorus*
5 a One pair of legs shorter than the other pairs	*Pulex*
b All pairs of legs of similar length	*Pediculus*

Table 9.1

	1 a	1 b	2 a	2 b	3 a	3 b	4 a	4 b	5 a	5 b	name of arthropod
A		✓					✓		✓		*Pediculus*
B											
C											
D											
E											
F											

[5]

[Total: 5]

Cambridge IGCSE Biology 0610 Paper 2 Q1 June 2009

10 a Sexual reproduction in flowering plants involves both pollination and fertilisation.

i Explain the difference between pollination and fertilisation.

....

....

....

....

.... [3]

ii Name the part of a flower where pollination happens.

.... [1]

iii Name the part of a flower where fertilisation happens.

.... [1]

b Sexual reproduction in flowers results in the production of seeds and fruits. From which part of a flower is each of these formed?

seed ....

fruit .... [2]

c Describe the role of the wind in the life cycle of some flowering plants.

....

....

.... [2]

[Total: 9]

Cambridge IGCSE Biology 0610 Paper 2 Q3 June 2009

11 Respiration is one of the characteristics of living things.

a List four other characteristics of living things **not** including respiration.

1 ....

2 ....

3 ....

4 .... [4]

b Describe the difference between *respiration* and *breathing*.

....

....

.... [2]

[Total: 6]

Cambridge IGCSE Biology 0610 Paper 2 Q1 June 2008

12 Fig. 12.1 shows the female reproductive system.

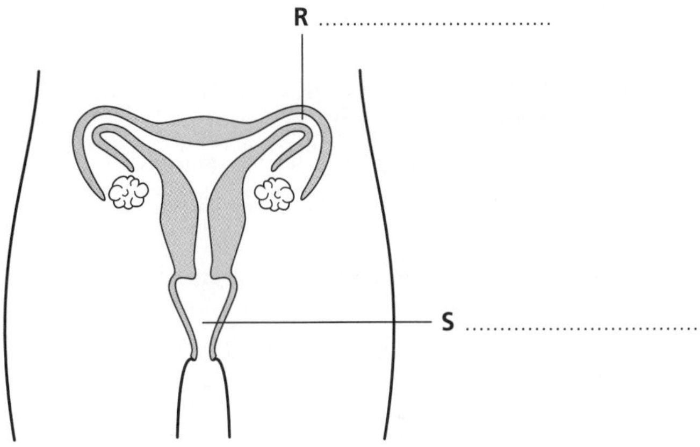

R ....

S ....

Fig. 12.1

a On Fig. 12.1, label structures **R** and **S**. [2]

b On Fig. 12.1,

i label, with a line and a letter **F**, where fertilisation occurs, [1]

ii label, with a line and a letter **I**, where implantation occurs. [1]

[Total: 4]

Cambridge IGCSE Biology 0610 Paper 2 Q6 June 2008 – not all parts of the question have been used

Acid rain Rain with a pH of less than 7, formed when an acidic gas dissolves in rainwater.

Active site The part of an enzyme that attaches to molecules and speeds up their reactions.

Active transport The use of energy to move substances across cell membranes.

Adaptation A physical feature or behaviour that suits plants and animals to their environment.

Addictive Describes a drug that changes brain cells so people need to keep taking it.

Aerobic respiration A reaction that uses oxygen to release energy from glucose.

Aeroponics Growing plants with their roots in air instead of soil.

Alcohol The substance produced when yeast ferments sugary solutions.

Algae Organisms found in water. They carry out photosynthesis but lack stems, roots, and leaves.

Alimentary canal The tube that carries food from the mouth to the rectum.

Alveoli Tiny air sacs in the lungs where gas exchange takes place.

Amino acids The type of molecule that makes up proteins.

Anaemia A deficiency disease in humans caused by a lack of iron.

Anaerobic respiration A reaction that produces energy without using oxygen.

Antagonistic Describes muscles that pull in opposite directions.

Anther The upper part of a stamen where pollen is produced.

Antibody A molecule made by white blood cells that helps to destroy pathogens.

Anus The end of the alimentary canal through which faeces exit the body.

Artery A thick-walled blood vessel that carries blood away from the heart.

Arthritis A painful disease that occurs when the cartilage at the ends of bones wears away.

Arthropods The most common type of invertebrate. They all have jointed legs.

Asthma A disease that makes breathing difficult because the bronchioles become too narrow.

Athlete's foot A fungal infection that occurs between the toes and causes the skin to become flaky.

Backbone A structure made up of vertebrae that runs from the base of the skull to the pelvis.

Ball-and-socket joint Joints that allow arms and legs to move in every direction.

Beetles Insects that have tough covers over their wings.

Beri-beri A deficiency disease in humans caused by a lack of vitamin B1.

Bile A substance released into the small intestine to emulsify fats.

Bioaccumulation The gradual build-up of persistent organic pollutants in food chains.

Biodiesel A biofuel made from plant oils.

Biodiversity The number and variety of species present.

Bioethanol A biofuel made out of sugars.

Biological catalysts Naturally occurring molecules (enzymes) that speed up chemical reactions.

Biomass The mass of living material in an area.

Biometrics Identifying people using unique characteristics such as fingerprints.

Brain The organ that takes in data from sense organs, takes decisions, and controls your actions.

Bronchiole Tiny tubes that carry air to the alveoli in lungs.

Bronchus (plural: bronchi) A tube that caries air from the trachea to one lung.

Budding A method of reproduction. Offspring develop as buds on the sides of parent cells.

Caffeine The drug found in coffee which is a stimulant.

Camouflage A colour or pattern that helps an organism to blend in with its surroundings.

Cancer A disease caused by cells dividing faster than they should.

Capillary A very thin-walled blood vessel used to deliver substances to cells.

Captive breeding programme Breeding endangered animals in protected environments such as zoos.

Carbohydrase The type of enzyme that breaks down carbohydrates.

Carbon monoxide A toxic gas that reduces the amount of oxygen red blood cells can carry.

Carbon-neutral Describes fuels that don't change the amount of carbon dioxide in the air.

Carcinogenic Describes a substance that can cause cancer by making cells divide faster than they should.

Carnivore A meat-eating animal.

Carpel The female part of a flower.

Cartilage A smooth tissue that covers the ends of bones and lets them slide over each other.

Catalysts A substance that speeds up chemical reactions.

Category A group that share characteristics.

Cell The building block that makes up living things.

Cell membrane The outer layer of a cell which controls what enters and leaves it.

Cell wall The tough outer covering of plant cells which helps support them.

CFC The type of compound used in early refrigerators which depleted the ozone layer.

Chemical digestion Breaking down large molecules into small molecules using enzymes.

Chlamydia A common sexually transmitted infection that can cause infertility.

Chloroplast A green structure found inside plant cells where photosynthesis occurs.

Chromosome Strings of genes made from DNA found in the nucleus of every cell.

Ciliated cell A cell with tiny hair-like structures that can sweep material along tubes.

Circulatory system The heart and blood vessels that transport blood around the body.

Classification Sorting organisms into groups based on their similarities and differences.

Climate change A change in the average weather conditions.

Cold A common infection caused by a virus which causes a runny nose, sore throat, and fever.

Community A group of species that live in the same place.

Conception The creation of life by the fertilisation of an egg cell by a male sex cell.

Condoms Devices that can protect people from sexually transmitted infections.

Conservation Saving living things from extinction and protecting their natural environment.

Consumer An organism that obtains food by eating other organisms.

Continuous variation Variation shown by characteristics that can take any value within a range like height.

Contract Get shorter.

Controlled The value of these variables must not change during an investigation.

Correlation A relationship between two variables.

Cytoplasm The gel-like fluid in cells where chemical reactions take place.

Decomposers Organisms that break down dead plants and animals and their waste products.

Deficiency diseases Diseases that occur when people don't get enough of an essential nutrient.

Deforestation The permanent removal of forests.

Denatured Describes an enzyme when its shape has changed so that it no longer works.

Depressant Type of drug that slows your reactions.

DNA The giant molecule that genes are made from.

Diaphragm A thick sheet of muscle below your lungs that contracts when you breathe in.

Differentiated Describes cells that have become specialised to do specific jobs.

Digestion The process that uses enzymes to break down the large molecules in food.

Digestive system The organs that digest and absorb food.

Discontinuous variation Variation shown by characteristics that can only take set values, such as blood groups.

Dislocate A bone moving out of its socket.

Dominant Describes a gene that always affects an organism even when only one copy is present.

Dormant Alive but not growing, for example a dry seed.

Dough A thick mixture of flour and liquid, used to make bread.

Ecosystem The living things in an area and the environment they interact with.

Egg cell A female sex cell (gamete).

Ejaculate The ejection of semen from a man's penis.

Electron microscope A microscope that uses an electron beam to produce a highly magnified image.

Embryo A new organism that is growing from a fertilised egg but does not have a full set of organs yet.

Emulsifies Breaks down into smaller droplets.

Endothermic Describes a reaction that takes in energy.

Environmental variation Variation between individuals caused by the environmental conditions they experience.

Enzyme Protein that speeds up reactions such as those that break down large food molecules.

Eutrophication The effect of adding too many minerals to a lake or river.

Evolution The way a species gradually changes as a result of survival of the fittest.

Evidence A collection of observations or measurements which may support an explanation.

Excretion The removal of waste products from the body.

Exothermic Describes a reaction that releases energy.

Explanation A possible answer to a question.

Faeces Solid waste excreted from your digestive system.

Ferment Break down sugar into carbon dioxide and alcohol.

Fertilisation Fusing the nucleus of a male sex cell with an egg cell to create a new life.

Fertilised Describes an egg when a sperm cell has fused with it.

Fertiliser A product used to improve plant growth by supplying extra minerals to the soil.

Fetal alcohol syndrome Birth defects caused by a mother drinking too much alcohol while pregnant.

Fibre A food component that cannot be digested but keeps the gut healthy.

Filament The lower part of a stamen in a flower, which supports the anther.

Flaccid Describes plant cells that have lost water and cannot support themselves.

Flowers These contain the reproductive organs of flowering plants. They produce seeds.

Flowering plants Plants that produce flowers and reproduce using seeds.

Flu A common infection caused by a virus which causes a runny nose and tiredness.

Flue gas desulfurisation A method used to remove sulfur dioxide from waste gases.

Food poisoning Vomiting and diarrhoea caused by eating food contaminated with microbes.

Food web Diagrams that show the connections between different food chains.

Fortified Describes food that has extra nutrients added to it.

Frequency chart Chart used to display discontinuous variation by showing the number in each category.

Fruit The structure that develops around a seed.

Fungus A type of micro-organism that absorbs nutrients from its surroundings.

Gall bladder The digestive organ that produces bile.

Gametes Male and female sex cells.

Gas exchange The process that takes place in your lungs when blood swaps carbon dioxide for oxygen.

Genes You inherit these from your parents and they influence the way your body develops.

Genetic engineering Adding new genes to cells.

Germination This happens when a seed starts to grow.

Global warming A gradual rise in the temperature of Earth's atmosphere and oceans.

Glucose A small carbohydrate molecule made by photosynthesis and broken down by respiration.

Gonorrhoea A common sexually transmitted infection that can cause infertility.

Greenhouse A structure that lets sunlight in but stops heat escaping.

Greenhouse effect The way some gases absorb heat and raises the temperature of Earth's atmosphere.

Growth factors Chemical signals released by cells that affect the way other cells develop.

Gut The tube that carries food from the mouth to the rectum.

Habitat The place where a plant or animal lives.

Haemoglobin A chemical that carries oxygen inside red blood cells.

Hallucinogen A type of drug that distorts your senses.

Heart The organ that pumps blood around your circulatory system.

Hepatitis C A serious disease caused by a virus that infects the liver.

Herbicide A substance that stops unwanted plants growing.

Herbivore A plant-eating animal.

Hinge joint The type of joint that allow arms and legs to bend and straighten.

Hormone A chemical messenger that affects cells all over the body.

Hybrid The offspring of two different species. They are usually infertile.

Hydroponics Growing plants with their roots in water instead of soil.

Hyperaccumulators Plants that can absorb large quantities of minerals.

Hyphae The thin threads that make up the body of a fungus.

Ice cores Columns of ice drilled from ice sheets.

Immune system The body system that identifies and destroys most pathogens.

Immune Resistant to a disease.

Immunised Made immune to a disease using a vaccine.

Implantation This takes place when an embryo settles into the wall of the uterus.

Inbreeding When animals breed with close relatives.

Infectious disease Disease caused by micro-organisms that can spread from person to person.

Infertile Describes an organism that is not able to reproduce.

Infertility Inability to reproduce.

Inherited variation Variation caused by inheriting different combinations of genes.

Insecticide A substance that kills insects.

Insects The most common type of arthropod. They all have six legs.

Insulator A substance such as fat or air that minimises heat loss by conduction.

Intercostal muscles Muscles between your ribs that help to create breathing movements.

Interdependent Describes populations that affect each other, such as predators and their prey.

Invasive species A species that outcompetes the existing plants or animals in a habitat.

Iodine A chemical used to detect starch. It turns starch dark blue.

Joint The place where two bones meet.

Kidneys Paired organs in the abdomen that clean blood and produce urine.

Kwashiorkor A deficiency disease caused by a lack of protein in the diet.

Lactic acid A molecule produced when bacteria respire using the sugars in milk.

Ladybirds A type of beetle with brightly coloured, rounded bodies.

Large intestine The organ that absorbs water to turn a mixture of fibre and bacteria into faeces.

Latin The language used to give each species a unique, two-part name.

Leaves The organs in a plant which absorb solar energy and make food.

Leishmaniasis A disease caused by protozoa carried by sandflies.

Lichens Living things made from fungi and algae that can make their own food.

Ligaments The bands of tissue that hold bones together at joints.

Line of best fit A line that comes closest to all of the points on a graph.

Living indicators Species that can be used to show how much pollution is present.

Magnify Make something appear bigger than it really is.

Malaria A disease carried by mosquitoes that causes fever and chills.

Malnutrition The result of an unbalanced diet.

Mark and recapture A method used to estimate the number of animals in a population.

Mechanical digestion Using teeth to break food into smaller pieces.

Menopause The time in a woman's life when her periods stop.

Menstrual cycle A repetitive sequence of events that prepares the womb to receive a fertilised egg.

Micro-organism Organisms that are too small to be seen without a microscope.

Microscope An instrument used to make magnified images.

Mitochondria The compartments inside cells where aerobic respiration takes place.

Model A simple way of representing more complicated structures or ideas.

Mucus A sticky fluid that traps microbes so that they can be swept out of your lungs.

Muscle The tissue that contracts and pulls on bones to produce movement.

Muscle fibre A single muscle cell.

Muscular system All the muscles that move bones.

Mutation A change in a gene.

Natural selection The way a population's environment selects which individuals survive.

Negative correlation A relationship in which one variable increases, while the other decreases.

Nervous system The brain, nerves, and sense organs that control your thoughts and actions.

Neutralise React with an acid and convert it to a neutral salt and water.

Nicotine An addictive drug found in cigarette smoke, that acts as a stimulant.

Night blindness A deficiency disease caused by a lack of vitamin A.

Non-renewable An energy resource that will run out and cannot be replaced.

Nucleus The part of a cell that contains genes and controls the cell's activity.

Nutrient A useful substance obtained from food which is needed in a balanced diet.

Obese Someone is obese if their body mass is much higher than average.

Optimum pH The pH at which enzymes work best.

Organic matter Material that came from living things and can be decomposed by microbes.

Organism A living thing.

Organ A collection of tissues that work together, e.g. the heart.

Ovary A female reproductive organ that produces sex cells.

Oviduct One of the tubes in females that carry egg cells from the ovary to the uterus.

Ovules These produce female sex cells inside the ovary of a flowering plant.

Oxides of nitrogen Acidic gases formed when petrol burns in a car engine.

Ozone A gas in the upper atmosphere that stops harmful ultraviolet radiation reaching Earth.

Palisade cell A plant cell found near the top of a leaf. Most photosynthesis takes place here.

Pancreas A digestive organ that releases enzymes into the small intestine.

Pasteurisation Heating food or drink to a temperature that kills microbes without ruining its flavour.

Pathogen A micro-organism that causes disease.

Peak expiratory flow (PEF) The maximum speed at which someone can expel air from their lungs.

Penis The male organ that deposits sperm in the vagina during sex.

Period The bleeding that occurs once a month in females.

Peristalsis The squeezing action the gut walls use to push food along.

Phagocyte A white blood cell that detects and destroys pathogens.

Pitfall trap A container buried in the ground used to trap invertebrates.

Pharming Using genetically engineered plants and animals to produce medicines.

Phloem A tube that carries sugars from leaves to other parts of a plant.

Photobioreactors Tubes surrounded by artificial light in which algae can be grown.

Photosynthesis The reaction green plants use to make their own food from carbon dioxide and water.

Phytoextraction Using plants to remove minerals from the soil.

Plasmid A small loop of genes found in bacteria.

Platelets Small fragments of cells that help blood to clot when a blood vessel is damaged.

Pollination Moving pollen from an anther to a stigma.

Pollution A harmful substance that has been added to the environment, e.g. fertiliser in rivers.

Population The number of individuals in a group of living things.

Positive correlation A relationship in which one variable increases, as the other increases.

Predator An animal that hunts other animals for food.

Prediction The expected outcome of an experiment based on scientific ideas.

Prey An animal that is hunted or killed for food.

Primary consumers Animals that feed on plants or other organisms which make their own food.

Producer An organism that can make its own food like a plant.

Protease The type of enzyme that breaks down proteins.

Protozoa A group of micro-organisms.

Puberty The time when young people's bodies change to prepare them for reproduction.

Pulse The regular throb of an artery that indicates your heart rate.

Quadrat A square frame used to find the number of plants growing in a known area.

Questions Things a scientist wants to find out.

Range The difference between the highest and lowest values in a sample.

Rectum The final part of the large intestine where faeces are stored.

Recessive Describes a gene that has no effect when only one copy is present.

Red blood cell Cells in the blood that carry oxygen around the body.

Rejected A transplanted organ or tissue that is not accepted by the recipient's body.

Renewable Describes an energy resource that will not run out because it is constantly replaced, e.g. wind power.

Repeatable Describes measurements that are the same every time they are taken.

Reproduction The production of offspring.

Resistant Describes individuals that survive in the presence of chemicals that are usually toxic.

Respiration The release of energy from glucose and oxygen in living things.

Respiratory system The organ system containing the lungs and the muscles that control them.

Ribs Curved bones that form a cage around the heart and lungs.

Rickets A deficiency disease caused by a lack of vitamin D or calcium.

Ringworm An infectious disease caused by a fungus growing in your skin.

Root hair cell Cells with long, thin extensions that increase the surface area of roots.

Roots Plant organs that hold the plant in the ground and absorb water and minerals.

Saliva A slippery fluid produced in the mouth which contains a carbohydrase enzyme.

Scaffold A support that gives soft tissue structure and makes it the right shape.

Scavenger An animal that feeds on dead plant or animal material.

Scientific question A question that involves one variable that can be changed and one that can be measured.

Scurvy A deficiency disease caused by a lack of vitamin C.

Secondary consumers Carnivores that feed on primary consumers.

Seed A structure made to protect the embryo when a flowering plant reproduces.

Selective breeding Choosing the parents of the next generation so they inherit specific characteristics.

Semen A fluid containing sperm.

Sense organs These detect changes in your environment and send nerve impulses to your brain.

Sensitivity The ability to detect changes in the environment.

Sewage Dirty water containing urine and faeces.

Sex hormone A chemical that produces the changes that take place during puberty.

Sexually transmitted infections (STIs) Diseases passed from one person to another through sex.

Skeleton The framework of bones and cartilage that supports an organism.

Skull The bone that surrounds and protects a vertebrate's brain.

Sleeping sickness A disease caused by protozoa carried by tsetse flies.

Small intestine The part of the digestive system where most nutrients are absorbed into the blood.

Smallpox An infectious disease that was wiped out using vaccines.

Solar cells Devices that transfer energy from light to electricity.

Specialised Suited to do a specific job.

Species A group of organisms with similar characteristics that can produce fertile offspring.

Specimen A section of tissue that can be observed under a microscope.

Sperm cell A male sex cell.

Sperm duct One of the tubes that carry sperm from the testes towards the penis.

Spinal cord The bundle of nerves inside your backbone.

Spongy mesophyll cells Plant cells near the bottom of leaves with air large spaces between them.

Spontaneous generation A theory that microbes are formed from decaying matter; the theory has been disproved.

Spores Reproductive cells made by moss, ferns, and fungi.

Sprain An injury to a joint caused by overstretching a ligament.

Stamens The male parts of flowers.

Starch A large carbohydrate molecule made from lots of glucose molecules joined together.

Stem cells Cells from an early embryo that can divide to produce any other sort of cell.

Stem The plant organ that holds up the leaves and transports food, water, and minerals.

Sterilise Destroy the micro-organisms in or on something.

Stigma The upper part of a carpel which catches pollen.

Stimulant A drug that make you feel alert or speeds up your reactions.

Stoma (plural: stomata) A gap on the underside of a leaf which allows gases to pass in and out.

Sulfur An element found in small amounts in fuel, which burns to create an acidic gas.

Sulfur dioxide An acidic gas formed when most fuels burn.

Survival of the fittest The survival of individuals that are best adapted to their environment.

Style The middle part of a carpel in a flower, which connects the stigma to the ovary.

Streamlined Describes a shape that is narrow at both ends to reduce friction.

Surface area The number of square metres covered.

Synovial fluid A slippery liquid found in most joints which provides lubrication.

Systems Groups of organs that work together to do a specific job.

Tar A mixture of chemicals in cigarette smoke that can paralyse cilia and cause cancer.

Tendon The strong tissue that connects muscle to bone.

Tertiary consumers Animals that feed on secondary consumers.

Tissue A group of similar cells that work together to perform a particular function.

Top predator The animal at the top of the food chain, that is not eaten by other animals.

Trachea The tube that carries air from your mouth and nose to your bronchi.

Transpiration The flow of water through a plant.

Transplant Move tissues or organs from one person to another.

Tree rings The rings of wood in a tree trunk.

Trophic level An organism's position in a food chain. Producers occupy trophic level 1.

Turgid Describes a firm, rigid cell full of water. Its vacuole pushes against the cell wall.

Typhoid An infectious disease spread by eating food or drink contaminated by bacteria.

Ultraviolet radiation A type of light that can damage plants and cause skin cancer in humans.

Urine Fluid excreted by the body that contains excess water and waste products.

Uterus The female organ where a baby develops.

Vaccine A weakened pathogen that trains your immune system to make antibodies to it.

Vacuole A fluid-filled sac in plant cells that helps the cell wall to provide support.

Vagina A tube from the outside of a female's body to the uterus.

Valid A conclusion that can be trusted because it is based on reliable results from fair tests.

Variable Something that can have more than one value.

Variation The differences between individuals.

Vector An organism that carries a pathogen from one person to another.

Vein A thin-walled blood vessel with valves that leads blood towards the heart.

Vertebrae Small bones that make up the backbone which surrounds and protects the spinal cord.

White blood cell Blood cells that can identify or attack pathogens.

Wildlife sanctuaries Special areas where animals can be protected.

Wilt What a plant does when its leaves and stems droop due to lack of water.

Xylem A tube that carries water and minerals from a plant's roots to its leaves.

Yeasts Microscopic single-celled fungi that do not have hyphae.